LIFTING THE VEIL

Lifting the Veil

THE FEMININE FACE OF SCIENCE

LINDA JEAN SHEPHERD, PH.D.

AN AUTHORS GUILD BACKINPRINT.COM EDITION

Lifting the Veil
The Feminine Face of Science

AN AUTHORS GUILD BACKINPRINT.COM EDITION
Published by iUniverse, Inc.

For information address:
iUniverse, Inc.
2021 Pine Lake Road, Suite 100
Lincoln, NE 68512
www.iuniverse.com

Because of the dynamic nature of the Internet, any Web addresses
or links contained in this book may have changed
since publication and may no longer be valid.

Originally published by Shambhala Press

ISBN-13: 978-0-595-45771-7

Printed in the United States of America

CONTENTS

ACKNOWLEDGMENTS

> Indeed, I do not forget that my voice is but one voice, my experience a mere drop in the sea, my knowledge no greater than the visual field in a microscope, my mind's eye a mirror that reflects a corner of the world, and my ideas—a subjective confession. —C. G. JUNG

This book is a personal journey, embarked upon to discover and honor the emerging Feminine in myself and in our culture. I am indebted to all those before me who stimulated my thinking, shaped my ideas, and changed my life through their writing. I wish to thank all those whose lives touched mine to help me form and nurture this book into reality.

I am grateful to the scientists whose stories are woven into the fabric of this book: Becca Dickstein, Marti Crouch, Sylvia Pollack, Sigrid Myrdal, Diane Horn, Edgar Mitchell, Davida Teller, Ingrith Deyrup-Olsen, Aimee Bakken, Kristina Katsaros, Paula Szkody, Patricia Thomas, Marsha Landolt, Sara Solla, Sue Kilham, and Eberhard Riedel—as well as those who requested anonymity. They generously made time for our interviews, and I appreciate their willingness to talk openly about painful as well as exciting parts of their work and lives. Thanks also go to Marti Crouch for sending me several articles, and to Brian Martin for sending me a copy of his book *The Bias of Science*.

The Women, Science, and Technology Reading and Discussion Group has provided an ongoing forum for exchanging ideas and perspectives. Thanks go to Viki Sonntag, Rebecca Hoff, Lisa Liebfried, Annemarie Succop, Julie Deans, Ann Grant, Michele Garfinkel, Chris Guenther, Gail Kineke, LuAnne Thompson, Ellen Gottheil, Chris Spengler, Katharine Grant, Christine Gibson, and

Acknowledgments

Lisa Leibfried for sharing their experiences in science. In particular, I wish to thank Rebecca Hoff for her selection of reading materials, Viki Sonntag for her article, reading materials, and ideas about technology policy, and Angela Ginorio for establishing forums for discussing research on women in science.

Thanks go to the organizers of the Puget Sound Science Writers Association—Paul Lowenberg, Kristine Moe, Bill Cannon, Tom Paulson, and Richard Strickland—for organizing events such as the panel discussion on "Whose Science Is It, Anyway?" and the demonstration of remote observing at the University of Washington to the Apache Point Observatory.

I continually learn about nurturing, receptivity, and generosity from Paul Hamilton, my partner in the alchemical dance of the Masculine and Feminine. Throughout the process he has lovingly supported my efforts with computer software and human emotionware. He shares my frustrations and elations, contributes his insights, and also edited early drafts.

I give heartfelt thanks to Anne de Vore for helping me to shore up my foundations psychologically, and to reclaim my feeling and the Feminine. Through her warmth, encouragement, and her belief in me and the work, I found the courage to speak my truth.

I am grateful to Ilene Stein for her nurturing and sustaining "thereness." By embodying the relatedness and connectedness of the Feminine, she has taught me their immeasurable value. Her thoughtful reading of many draft chapters and our ongoing conversations have significantly improved this work. Thanks go also to her husband, Jim Oswald, for his contributions to our discussions.

My friends and scientific colleagues Diane Horn and Becca Dickstein brought new material to my attention and took time from their busy lives to review parts of the book. Diane graciously hosted a group of cell biologists in the comfort of her living room as we explored the role of the Feminine in their work.

For their patience, encouragement, good humor, and stimulating discussions, I wish to thank the members of our writers' critique group: Marie Landis, Bruce Taylor, Phyllis Lambert, Calvin Clawson, and Brian Herbert. Each gave their criticism in the spirit of love and helpfulness during our monthly gatherings. Thanks go to Claudia McCormick for introducing me to this productive group of writers.

I am deeply grateful for the support of my friends and colleagues. Paul-Erik Jensen started me thinking about issues of the Feminine. Deanne Jameson loaned me books, reviewed several chapters, and shared her thoughts about how science has influenced her concepts of reality. Sylvia Pollack critiqued an early article on the subject and inspired me to expand the scope of my research. Aimee Bakken graciously opened her lab to me. Jan Alt ironed out the kinks in my body acquired from hours at the word processor. Brenda Bentz provided a haven in her Guest Nest on the Hood Canal, where I was enlivened by our conversations about feminism and the value of "thinking like a woman." Science writer Joel Davis generously provided guidance as I was getting started as a writer. Artist Patianne Stevenson executed a working drawing of the image I envisioned on the cover of the book. My literary agent, Natasha Kern, started me on the path of expanding my ideas into a proposal, championed the project, and guided me through the process of finding a publisher.

I extend my deepest appreciation to Emily Hilburn Sell at Shambhala Publications for her sensitive and skillful editing. Always a pleasure to work with, she nurtured the book through her feedback and enthusiasm for the project—and by sharing my belief in the importance of the Feminine.

The staff of the Issaquah Library has been helpful in locating books and reference materials available through the wonderfully convenient interlibrary-loan system of the King County Library. Although it occurred long in the past, I wish to acknowledge the American Association of University Women for awarding me a fellowship to support me during my last year of college.

Finally, I wish to thank all those who have sustained me with their friendship, love, and encouragement: my father Don Shepherd, my brother Gregg Shepherd, Marla Herbig, Ruth Pickering, Denis Janky, Janet Colli, Pat Lantow, Bill Field, Brad Fisher, Mary Dunbar, Tavi Karpilow, Ruth Pickering, Basil Bumagin, Paul Noll, and Marsha Hayes Merz.

PREFACE

I learned early in my life what it meant to be a scientist. Only in the last seven years have I struggled consciously with what it means to be a woman—and a woman in science.

I entered the realm of science during the late 1960s, just as the women's liberation movement was organizing. To me, feminist meant the extremists who burned their bras while guys were burning their draft cards. Whereas burning draft cards seemed an important political statement, burning bras just seemed silly. And I could not understand all that squabbling over *Ms.*

In my teens I was eager to find out whom I was going to marry so I could align my life and career with his—working at his side and sharing his interests. Since I enjoyed all subjects in school and got mostly A's, my adolescent bravado told me I could do anything. Then I started dating John. He spent evenings, weekends, and summers in his basement laboratory. In his chemistry lab he strove to re-create the primordial conditions on Earth that led to the synthesis of nucleic acids, precursors of life. Amidst a maze of outdated electronics gear, he and his friend Tom planned to attract a flying saucer with their Tesla coil, a sewer pipe wrapped with pink and white wire that generated a huge arc of electricity.

My idea of being feminine meant wearing ribbons and frilly blouses, and pursuing quiet activities like cooking and sewing. Although I became proficient at these activities, they seemed humdrum, insignificant, and devoid of meaning. In fact, nothing women did—mothering, teaching, nursing, being a secretary—seemed valued. Science, on the other hand, sparkled with power.

Along with most of my generation, I grew up embedded in the scientific framework. Without realizing it, my reality was defined by science. I believed that if science had not proven something, then it did not exist. I had faith that, given enough time and money, scientists could solve all of our problems and answer all of our questions. After all, scientists had landed a man on the moon.

John and I attended college together, married, and entered graduate school to study biochemistry. It never occurred to me that anything feminine could be relevant to science. Science is, after all, an objective and rational discipline. It is the scientific method, a rigorous way of obtaining knowledge. What possible roles could feeling and nurturing have? How could love have anything to do with science? Either a theory is validated and accepted, or it is disproved by further experiments.

During the seventies, I enjoyed the incongruity of being a woman in science. But my conscious expression of anything feminine was limited to occasionally wearing dresses and stockings, having long hair, making tea and sandwiches for John, and cleaning his glassware. Through lectures and lab work, surrounded by other rational scientists, I became even more inculcated into the scientific worldview. Sometimes things didn't make sense or certain assumptions didn't seem right, but I thought the problem was with me—that I didn't know enough—so I didn't question their validity.

Along with many other professional women of my generation, I saw that power in our culture was aligned with the things that men did—science, business, law, politics. In order to prove myself and succeed in the male realm of science I adopted the rational, analytical, hierarchical approach. I wanted to prove that I could be just as smart and competent as men.

Not until seven years ago did I seriously question what it meant to me to be a woman. John and I had divorced and I had no children, so the traditional roles of wife and mother did not apply to me. Although I made more money as a scientist in a biotechnology company than I had imagined possible, my job felt sterile and did not satisfy me. I felt exhausted. I yearned to do work that had more meaning for me, something that would make a difference.

Then I discovered the richness of the Jungian idea of the "feminine principle" as a fundamental force in our psyches. As conceived by C. G. Jung and amplified by modern Jungians, the Feminine (with a capital *F*, to distinguish the archetypal feminine principle from the superficial sweetness-and-spice-and-everything-nice notion of feminine) describes the archetypal force of relatedness carried primarily by women in our culture. This is the force that attracts, connects, and holds people together. According to Jung, as people journey toward wholeness, men integrate the feminine side of themselves and women integrate the masculine side of themselves. For most people, this process of integration begins to occur at midlife. For me, it began in my twenties, but at the cost of denying the Feminine. This book is part of my personal journey toward revaluing and reclaiming the Feminine.

In reviewing this book for publication, several editors were intrigued by the idea but asked for a change in language away from the Feminine/Masculine dichotomy to avoid the emotional baggage that people bring to the word *feminine*. One editor suggested that I transcend the polarity with an androgynous term that included both. I agonized over this language issue for several weeks. I observed that other authors have confronted this same issue and resorted to less emotionally charged terms, such as left brain/right brain, or the Chinese concepts of *yin* and *yang*. I finally came to the conclusion that we cannot transcend this polarity until we equally value both parts—otherwise the Feminine will continue to be denied.

The danger of using neutral language for qualities that have been classified as feminine is that they are liable to be appropriated by men rather than accredited to women as their carriers. For example, during the nineteenth century it was claimed that women could not do science because they were not analytical enough. Now that scientists have discovered the value of a more intuitive approach, it is being said that women are too rational and cannot make creative intuitive leaps![1] Such appropriation leaves women in the same inferior position in a hierarchical power structure. While I am delighted when I see men are embracing qualities that have been carried by women for centuries, I am angered when men co-opt them and again exclude women from participation.

If these are the qualities that have been classified as feminine in our culture and have been primarily carried by women, then valuing these qualities should also reflect the value of the carrier. Doing so can help us be whole people. Carrying through the implications of equally valuing the Feminine would lead to a radical reframing of our concepts of science, how we see the world, and how we lead our everyday lives. While science has been a masculine endeavor, science in the broader sense of a search for knowledge and truth has no gender. I deeply feel that we can no longer afford to limit our search to such a one-sided approach.

Another puzzling reaction to the book came when one woman editor enthusiastically showed the proposal for this book to the head of the publishing company. He rejected it because he considered it sexist. I think he interpreted my honoring of the Feminine as automatically denigrating men, ignoring my emphatically stated desire to *equally* value the Feminine and Masculine. It seems hard for men who are vested with power to embrace qualities that our culture has viewed as powerless. Defensive, they quickly deny these parts of themselves and retreat to a macho male stance. In addition, many men shy away from exploring the Feminine in themselves because it gets mixed up with fears of homosexuality.

When we are enmeshed in such a hierarchical framework we automatically rank one person, profession, race, or gender over another. In doing so we fail to value the wonder, beauty, and benefits of diversity. While lab work grinds to a halt when the dishwasher gets sick, experiments can continue in the absence of the principal investigator for weeks at a time. A company or project will not succeed unless all aspects are equally well done—yet we continue to rank some people as more important to the endeavor than others.

As long as this hierarchical worldview predominates, being different from the white male professional means being inferior. Either you rank as First or Number One or you are dismissed as unimportant. Because of this, I have found that many women— even feminists—are nervous about identifying with anything feminine, since they have worked so hard to prove their equality with men. As a result they deny the parts of themselves that are different from men and are reluctant to explore any quality that could be labeled inferior, such as feeling or nurturing. Even to contemplate the notion that women may be different from men is threatening. For example, when I mentioned the topic of this book to a feminist historian of science, she looked away uneasily. I asked her why the word "feminine" made her so nervous, and she replied, "Because of biological determinism." In other words, if feminine characteristics are biologically determined, women are doomed to be forever inferior in a hierarchical world. On the other hand, a worldview that delights in diversity and sees differences as complementary, equally valuable, and beneficial to the whole, allows differences to be embraced and celebrated. An openness to multiplicity is critical to the reading of this book. For this reason, I ask the reader to experimentally suspend thoughts of automatically ranking one thing or person above another, and consider how a world would look that values the creative possibilities inherent in diversity.

LIFTING THE VEIL

1
VEILING THE FEMININE FACE
OF SCIENCE

Over the last fifty years, the magnificent achievements of science and technology have culminated in disastrous unforeseen consequences, tearing the very fabric of nature. The physics that landed a man on the moon also produced a world haunted by the threat of nuclear war. The chemistry that developed an incredible diversity of plastics also bequeaths a legacy of waste products that nature cannot reabsorb. The biology that brought forth the green revolution through fertilizers, herbicides, and pesticides threatens to yield a silent spring.

While I feel disappointed and fearful about how we choose to use the power of science, I also feel respect for that power and awe for our intellectual achievements. I care deeply about what happens to both humankind and the other species on this planet. When I read about the latest ecological disaster, I feel pain in the core of my being. I also feel the pain of friends and colleagues trying to survive in uncaring, unrelated institutions of science. I write out of hope that changes in individuals' ways of thinking about priorities, goals, consequences, and the very process of science can transform science into a force for freedom, truth, and creativity for all beings. I believe that the Feminine in each of us—the part of us that sees life in context, the interconnectedness of everything, and the consequences of our actions on future generations—can help heal the wounds of our planet. For these reasons I left experimental science and devoted four years of my life to this book.

Many people are troubled about the impact of technology on the environment. Some hold science responsible for the damage to our planet. Others see science as a tool to be used for good or ill, depending upon the values of the people wielding the tool. In addition, the majority of Americans feel intimidated by science and consequently shy away from participating in it. The number of American students entering science is declining. Many turn away from science not because it is difficult, but because it seems dry, dull, and unrelated to their lives. Each year the gap between scientists and the public widens, as researchers speak an increasingly separate language. New words and acronyms describe smaller and smaller bits of nature: quarks and bosons, exons and introns, HIV and AZT. This process of analysis and logic, reducing nature to its component parts, is characteristic of the masculine approach that has defined science. Until recently, most scientists considered feminine qualities, such as feeling and caring, irrelevant—if not detrimental—to their work. While we need the language to describe all the bits of nature, we also need to bring them into relationship with a broader perspective.

Science touches the lives of everyone on the planet. Electricity, automobiles, radios, computers, plastics, pesticides, gunpowder, and antibiotics are all products of science. Because of its success, many look to science to answer all their questions about the world. There is no doubt the grand adventure of science has provided remarkable knowledge about the universe and produced wondrous tools for the improvement of the human condition. Yet, caught up in the tremendous success of the objective, reductionistic approach, other paths to learning about nature have been neglected or suppressed. Examining the history of Western science reveals the intentional repression of one such approach, that representing the feminine viewpoint, which has been ignored from the outset.

WESTERN NOTIONS OF MASCULINE AND FEMININE

Are men and women really different? There are so many levels to this question that they are easily confused. Although considerable anthropological and sociological research by Margaret Mead and others indicates that gender is a cultural construct, sociobiologists such as E. O. Wilson have renewed the argument that behavior patterns typical of males and females are biologically determined. Between each side of the nature versus nurture controversy lies a vast gray area and a host of unanswered questions. For example, how tightly are intellectual, emotional, and psychological characteristics tied to gender? Intellectually, many women scientists and mathematicians have been recognized for their accomplishments. Emotionally, many male artists, musicians, and writers have shown that the Y chromosome does not render them incapable of great feeling. And the degree to which masculine and feminine psychological qualities are determined by biology or culture remains a hotly debated topic. Regardless of the origin of the differences, however, most cultures associate particular qualities with one sex or the other.

The writings of Aristotle (384–322 B.C.) reflected the thinking of his time and dominated Western thought for over two thousand years. The greatest collector and organizer of knowledge of the ancient world, Aristotle provided the only systematic survey of knowledge until the Renaissance. How did he view the world? What did he write about women? Did he value the Feminine?

Aristotle's concepts about women were derived from a cosmology based on observation and reason. He believed that order is pervasive and exists in increasingly subtle and complex hierarchies. Since "generation" and "corruption" were not observed in the heavens, he deemed the celestial region to be eternal and immutable. Reason and purpose reached perfection in the divine heavens, the abode of the gods. The Earth, on the other hand, had no such permanence. Earthly generation and corruption were clearly observable: the seasons came and went, animals were born, matured, reproduced, and died.

Aristotle applied the terms *male* and *female* to the cosmos. He spoke of the nature of the Earth as something female and called it "mother," while referring to heaven and the sun as "generator" and "father."[1] He maintained that whatever is superior should be separated as far as possible from what is inferior, thus explaining why the heavens are separate from the lowly Earth. Because the male possessed the superior faculties of reason and deliberation, it followed that "the relation of male to female is naturally that of the superior to the inferior—of the ruling to the ruled."[2] Aristotle considered femaleness "a deformity, though one that occurs in the ordinary course of nature."[3]

In the process of reproduction, Aristotle reasoned, the man's seed provides the active principle and the rational soul, while the woman, who is basically an infertile man with an animal soul, contributes merely the matter on which the active principle works. If all goes well, the sexual union produces a male offspring. If, however, the active principle is defective and does not overcome the resistance of the matter supplied by the female, then a female offspring results.[4] Aristotle wrote:

> Just as the young of mutilated parents are sometimes born mutilated and sometimes not, so also the young born of a female are sometimes female and sometimes male instead. For the female is, as it were, a mutilated male, and the catamenia [menstrual discharge] are semen, only not pure; for there is only one thing they have not in them, the principle of soul.[5]

In recent centuries, the soul has lost value, taken on feminine connotations, and has been projected onto women.[6]

Even today, many men experience femaleness as wholly alien and "other." As reproductive beings, women embody the natural, the disordered, and the irrational. Women often do not make sense to men; they seem mysterious, an enigma. In addition, a woman may arouse in a man confusing emotions and passions—feelings he may deem inconvenient. Love and hate, joy and sorrow, fear and rage, shame and guilt, influence a person's behavior in complex, often unpredictable ways, creating disorder and chaos. They interfere with clear, efficient, exact thinking. In order to make their way in the world, historically men have projected many disruptive and undesirable qualities onto their apparent source, women, denying any origin within themselves. In this way, men labeled as feminine those qualities they observed in women, together with those qualities they rejected in themselves and projected onto women. Similarly, "undesirable" characteristics also have been projected onto people of other races, nationalities, ages, and religions.

In Western culture, the successful man is considered to be objective, intelligent, logical, active, rational, independent, forceful, risk-taking, courageous, aggressive, competitive, innovative, and emotionally controlled. These qualities have been highly valued in our culture and are well rewarded financially. If a man is too soft or sensitive, he is labeled effeminate or womanish. He feels insulted if told he "thinks like a woman."

Western society expects women to be nurturing, receptive, passive, emotional, irrational, intuitive, subjective, compassionate, sensitive, kind, unaggressive, and uncompetitive. The positive value of these qualities has been minimized and dismissed as unimportant. In our materialistic culture, love seems irrelevant to the bottom line. This is evident in the fact that social service jobs, as labors of caring, are typically low-paying.

As in other masculine realms such as business and law, feminine qualities have been devalued and repressed in Western science. To many, the terms "feminine" and "science" are still mutually exclusive—scientists are presumed to be men. Publicly, many women scientists still say, "Oh, we don't think there is anything special about women in science. Science is science." They dare not talk about differences between the sexes, and try to convince others that differences do not exist. Many scientists are reluctant to express feminine qualities in their work for fear of losing credibility. A woman zoologist admitted:

> I am embarrassed to own up to the political sense I have, that if a certain kind of science is feminine or feminized, then somehow it's second class. And the upshot of that is that I tend to deny to all and sundry that my science has any taint of the feminine. I do that not only on my own behalf, but on behalf of what I still consider to be a beleaguered community of women scientists. . . . Let's not call it a special feminine way of looking at science, because I think that denigrates it at the present time.[7]

Unfortunately, we are equally ruled by the stereotypes of "feminine" behavior when we react to them and go to the other extreme of excluding them from our repertoire—thus denying them expression—as when we conform to the stereotypes.

Beliefs about what is masculine or feminine, about how men and

5

women should behave, act to constrain us within certain limits. These expectations vary from one culture to another. Exceptional individuals throughout history have challenged these limits, and questioned the assumptions underlying the restraints on their behavior. Nobel laureates such as Marie Curie, Irène Joliot-Curie, Maria Goeppert Mayer, Gerty Cori, Dorothy Hodgkin, Rosalyn Yalow, Barbara McClintock, Rita Levi-Montalcini, and Gertrude Elion, have proven that women are capable of the "masculine" virtues of intelligence and clear thinking. Yet, the collective belief that masculine and feminine qualities are exclusively tied to one sex or the other retains power over the majority within our culture. Desire to be accepted by peers is a powerful motivating force for staying within the limits of "normal." Many young girls still eschew math and science as "unfeminine," or are told by teachers that they cannot understand these subjects. One woman recalled her experience in high school physics:

> I remember answering the teacher's question about how a plane worked. It was something I understood pretty well. After hearing my answer, the class was quiet for a minute, and then the teacher looked at the boys in the back row and said, "Don't worry, girls only memorize those things, she doesn't really understand it."

Although 90 percent of her professors were supportive, one engineer remembered feeling invisible in some of her classes: "In college there were times when the professor asked a question and I was the only one who raised my hand with the answer. His eyes would sweep around the room and he'd announce, 'Well, I see that *nobody* has the answer.' Just unbelievable. About one out of ten professors was like that."

I think most men behave like this not out of malice, but out of ignorance or unconsciousness. For example, I recently consulted the local grange about testing our soil at home. The salesman gave me a form and in a friendly way suggested I take it home so that my husband could fill it out. Now, I am the gardener and chemist in the family and know volumes more about such things as pH than my partner—but it simply *did not occur* to the salesman that a woman could be knowledgeable and competent in this subject. In many ways, we are still lifting the veils of collective denial.

Individual by individual, we now must see past the limits of the mass mind and envision the possibilities that all of us unconsciously agreed to deny.

CROSS-CULTURAL PERSPECTIVES OF MASCULINE AND FEMININE

The concepts of masculine as the superior, celestial male, uniquely capable of action, reason, and deliberation, and the inferior, earthly female representing passivity, matter, and corruption are by no means universal. Egyptian cosmology, for example, reversed many of the characteristics we now associate with masculine and feminine. The sky goddess Nut overarched her husband, Geb, the earth god, portrayed lying on his back with his erect penis reaching toward Nut. When the two intermingled, Creation arose. Maat, the principle of order, truth, law, justice, and harmony, manifested as a goddess and was associated with the celestial harmonization of the realm of generation. Chaos, disorder, and destruction were represented by a male god, Typhon, or Set. The god Osiris was the deity of death, lord of the flood and vegetation. His wife Isis played the active role in Egyptian mythology by searching for the scattered pieces of Osiris's dismembered body.

Although separate deities represented heaven and earth, the ancient Egyptians did not sever the earth from the sky; there was no duality between humanity and nature. The images of Nut and Geb, for example, portrayed their conjugal relationship. Divine Reality manifested in every aspect of the creation. For the Egyptians, humanity as microcosmos mirrored the macrocosm: all within reflected that which was without; the essences of humankind and the essences of the cosmos were one; as above, so below.

In more modern times, the anthropologist Margaret Mead reported on cultures that defied the sex roles imposed on Western culture. In particular, she studied the Tchambuli tribe in New Guinea where the usual patterns of behavior of men and women were reversed. Here women managed the business affairs, earned the money, fished, gardened, traded, and cooperated well with each other, while the men carved and painted, adorned themselves, gossiped, were moody, and engaged in ceaseless petty jealousies

7

and rivalries with other men. Mead concluded that the stereotypic feminine and masculine traits were not inborn but were the result of cultural conditioning. The way children were reared, the way certain kinds of behavior were rewarded or punished, the way heroes and villains were portrayed—those elements, and not innate characteristics, influenced the development of a child's temperament and personality. Furthermore, the way in which a society was organized determined the roles males and females were expected to play and therefore what talents and temperaments would be developed in children of the respective sexes.[8]

Whereas Aristotle placed value judgments on feminine and masculine qualities, other cultures devised systems of classification that were not value-laden in this way. Chinese Taoist scientists, for example, describe the concept that all of life is comprised of, and has been set in motion by, the constant interplay of two vital energies: the principles of the passive yin (everything female) and the active yang (everything male). Together, yin and yang give rise to everything in the world—the "ten thousand things."

In this system, no part has a life of its own, but each exists in complementary interaction with the other. Yin and yang mutually help each other; together they constitute equilibrium and harmony. Though opposite, they are not in opposition or antagonistic. Though different, they supplement each other. There is a continuous movement between them, without beginning and without end: when yang reaches its final moment, then yin is manifested; when yin is completed, yang begins again. Instead of representing a hierarchy of superior and inferior qualities, they represent a duality that exists in harmonious relationship, as necessary parts of the whole.

The symbol of this relationship is a circle divided into two curved shapes of equal size, the dark right part signifying yin and the light left part representing yang. Within yin is a spot of yang, and within yang is a dot of yin, showing that each contains an element of the other. The line between them has the movement of a wave, representing the continuity of the life force, which is movement. This is very different from Aristotle's concept that the divine is immutable and ungenerative, and that change is corrupt.

YANG	YIN
Primal power, energy	Yielding
Thrusting, active, dynamic	Receptive, passive
Thinking	Feeling
Logos, the principle of objective interest, of fact and logic	Eros, the principle of relatedness
Knowledge for the sake of knowledge	Application of knowledge
Analytic	Comprehensive
Disciplinary	Nondisciplinary
Order	Chaos
Accomplishment, efficiency	Pleasure, enjoyment
Experiment, adventure	Security, familiarity
Competition	Sense of community
Focused consciousness	Diffuse awareness
Head, intelligence	Soul, body
Information obtained through the senses	Intuition
Tangible	Intangible
Concentration, determination	Relaxation
Pursuit, construction	Submission, maintenance
Hard	Soft
Fire, air	Earth, water
Dry	Moist
Light, day	Dark, night
Sun	Moon
Hot	Cold
Summer	Winter
Positive	Negative
Vertical	Horizontal
Meat	Vegetables

From its inception, Chinese science included qualities defined by the West as feminine. Rather than aspiring to dominate nature, the Chinese regarded an intimate and harmonious relationship with the natural forces as the ideal:

> The universe did not exist specifically to satisfy humans. Their role in the universe was to "assist in the transforming and nourishing process of heaven and earth," and this is why it was so often said that humanity formed a triad with heaven and earth. It was not for man to question Heaven nor compete with it, but rather to fall in with it while satisfying his basic necessities. . . . Hence the key word is always "harmony"; the ancient Chinese sought for order and harmony throughout natural phenomena, and took this to be the ideal in all human relationship. Thus, for the Chinese, the natural world was not something hostile or evil, which had to be perpetually subdued by willpower and brute force, but something much more like the greatest of all living organisms, the governing principles of which had to be understood so that life could be lived in harmony with it.[9]

Within China, the Taoists were the ones primarily responsible for the advancement of experimental science. While no Confucian scholar would sully himself with manual work of any kind, the Taoist considered experimental science part of his search for the Tao.

Chinese science never separated spirit and matter, or mind and body. Their physics eschewed mechanical materialism and physical reductionism, and remained faithful to their prototypic wave theory of yin and yang and the five elements (water, fire, wood, metal, and earth). These fundamental ideas never lent themselves to reductionism because the constituent elements were always inextricably linked together in the continuum. Yin and yang could not be separated out, isolated, or purified, even in theory. Abstract knowledge of the either A or not-A logic was avoided in Chinese science in favor of nonexclusive relationships of forces. The Taoists valued strength without hardness, energy without tenseness, vitality without nervousness.

Examination of traditional Chinese science indicates that the Western tendency to overexert technology in dominating nature, violently expending energy in deforestation, mining, and chemical-based agriculture is not the only way to use the magnificent power afforded by science. Chinese traditional medicine integrates spirit, mind, and body, diet and dreams, energy flows, and physical sensations, and remains empirically successful while resisting all efforts to define it within the categories of physiological reductionism. For example, Western science has not been able to account for the efficacy of acupuncture.

Unfortunately, Chinese science was dominated by a bureaucratic elite that resisted the introduction of new science and technology unless it provided clear benefit to the bureaucracy. Once China had achieved a condition of comparable social stability, the Chinese mandarinate did little to encourage innovation or trading contacts with the outside world. As a result, China dropped behind the West in technological skills, and fell to the West's military might.

THE JUNGIAN MODEL OF MASCULINE AND FEMININE

A somewhat more fluid, although culture-bound, definition of masculine and feminine comes from the depth psychology of Jung, who described two abstract patterns of human behavior *not* intrinsically allied with anatomical gender: Eros, the feminine principle of relatedness, and Logos, the masculine principle of objective interest. In regarding them as principles of human behavior, Jung saw Eros and Logos operating in the psyche as eternal opposites. He associated these qualities with Eros: emotionality, aesthetics, spirituality, value reached through feeling, subtlety, the urge to relate, to value, to join, to be in the midst of, to reach out to, to get in touch with, to connect, to get involved with concrete feelings, things, and people rather than to abstract and theorize. This sense of relatedness demands that we adapt our needs and desires to those of the other person and to the requirements of the situation. To do so we cannot remain in a fixed attitude but must be flexible.

With Logos, Jung associated reason, clear thinking, activity, high-mindedness, problem-solving, discrimination, judgment, insight, abstraction, and nonpersonal truth reached through objectivity. Because he had observed the crippling effects of one-sided development of one principle or the other, Jung insisted on the importance of Eros-relatedness for men and Logos-directness for

11

women. He observed that as an individual grows too one-sided, he becomes hindered, crippled, and loses flexibility:

> After the middle of life, permanent loss of the anima [the feminine principle in a man] means a diminution of vitality, of flexibility, and of human kindness. The result, as a rule, is premature rigidity, crustiness, stereotypy, fanatical one-sidedness, obstinacy, pedantry, or else resignation, weariness, sloppiness, irresponsibility. . . .[10]

Jung believed that our main task in life, the process of psychological growth toward wholeness, requires integrating the masculine and feminine principles. He stressed that each individual achieves wholeness only by developing and integrating both sides of themselves. This gives the individual more choices and resources for interacting with the world. Interestingly, when a man develops qualities of relatedness such as nurturing and receptivity, rather than seeming feminine, he becomes more intensely masculine—but without the defensive macho brittleness. In the same way, when a woman integrates qualities of discrimination and clear thinking, she can do so in a deeply feminine way, where masculine reason is tempered with compassion.

In the individual, exclusive reliance on either masculine or feminine consciousness also inhibits the development of that consciousness itself, since neither principle can realize its full potential without continual reference to its opposite. Psychotherapist and writer Sukie Colgrave observes that a person who is excessively controlled by the masculine discriminating principle may feel that life has lost its meaning.[11] Oscillating between arrogance and despair, such a person experiences a growing alienation from other people and from the Self. The Masculine continues to differentiate unaided by the complementary influence of the Feminine, which functions to reveal the connections and relationships. Without relatedness, the person feels arid and loses a sense of direction in life. At this point, only the feminine attributes of listening, yielding, accepting, waiting, and trusting can revitalize the psyche. The Feminine creates the necessary relationships to inner and outer nature that restore meaning and purpose to life.

Conversely, excessive reliance on the feminine principle drags consciousness into a sea of sameness in which all differences are submerged. The person loses the capacity or will to act and think as an individual. The masculine principle is necessary to be able to focus and understand the different feminine qualities. It gives the person the ability to discriminate and to create a sense of an independent self, and is essential for intellectual understanding. Each principle sees the world through different lenses and therefore perceives different worlds. Developing both modes of consciousness enables a person to create a richer, more complete picture. Through long years of training and work, women scientists have assimilated the "masculine" capacities for discrimination, analysis, rational thinking, and abstraction necessary to do science. Sometimes the Feminine is lost in the process, because science defines itself in such masculine terms that the Feminine seems irrelevant to the objective experimental process.

While biological determinism implies that male and female attributes are hardwired, ruled by genes and hormones, an evolutionary perspective of the development of consciousness implies that these are human traits, accessible to us all. Although qualities may develop in a different sequence or be expressed in a different way by women and men, the goal of each person is development of our full human potential.

Now let us look at the Feminine and Masculine within the context of the Jungian theory of the evolution of consciousness. I believe this perspective can give meaning to historical developments (as opposed to assigning blame), and lift the debate above strictly gender related issues. In the same way that Jung and the Taoists view the Masculine and Feminine, yin and yang, I will refer to these principles not as rigidly framed definitions, but rather as fluid, flexible containers for certain qualities that both women and men possess.

PSYCHOLOGICAL DEVELOPMENT AND THE EVOLUTION
OF CONSCIOUSNESS

Just as human embryos pass through a developmental stage with rudimentary gill structures,[12] the development of consciousness of the individual recapitulates the development of consciousness of the culture. Naturally, this development embraces a wide spectrum

13

of individual variation. While some individuals may fail to manifest the full potential of consciousness exhibited by their culture, others may be the pioneers of new developments in the ongoing evolution of consciousness.

Erich Neumann's book *The Origins and History of Consciousness* outlines the stages in the development of consciousness based on the Jungian theory. Drawing on the evolution of mythological images, Neumann discusses how the history of Western consciousness parallels the development of consciousness in the individual. For our purposes, we will only briefly review the major stages in this course of development.

In the beginning was perfection and wholeness—the Garden of Eden, Humpty Dumpty before the great fall. The uroborus, the serpent coiled into a circle biting its own tail, symbolizes this stage that exists before delineation and separation of the opposites.

This unconscious wholeness corresponds with the early childhood stage of *participation mystique,* where individuals cannot distinguish themselves from the thing and feel at one with the object. The ego is contained in the unconscious; selfhood is identified with external objects and events. The individuality of the person has not yet developed. Mother, father, home, and the family dog are all part of the "I" of the child. When the child loses a cherished teddy bear, it feels like a loss of self. In this stage, the psyche knows a profound sense of interconnectedness, but only because it cannot tell the difference between self and other. All the opposites intermingle. Full of wonder, the child envisions limitless possibilities.

Gradually the child becomes aware of the mother as a separate object, while still being dependent on her. Psychohistorically, this stage of dependence on the mother correlates with cultures that worshipped the Goddess. A peaceful, passive, contemplative attitude prevailed that allowed the Great Mother and the primeval forces to do their work. Cultures at these early levels of psychological development practiced imitative magic, such as copulating in the cornfields to increase the fertility of the earth. The hunter exerted a magical influence on an animal whose image he possessed. Everything in the world was pregnant with meaning. In this matriarchal stage, the Feminine prevailed over the Masculine, the unconscious over the ego and consciousness. While the Goddess held sway in these remote historical periods, the Feminine was out in the open and prominent. The fundamental symbolic equation of the Feminine, woman = body = vessel included the world in this formula, woman = body = vessel = world.[13] But this prehistorical matriarchal phase lacked consciousness, awareness of itself. In his book *The Origin of Consciousness and the Breakdown of the Bicameral Mind,* Julian Jaynes shows how ancient peoples could not "think" as we do today, and were therefore not yet conscious.[14]

Next, the ego separates from maternal consciousness and moves toward independence and individuality. By developing analytical objectivity, the ego detaches from the natural context of the surrounding world and the unconscious. Historically, as *participation mystique* retreated, the nymphs and fauns disappeared from the groves. God and the angels moved to the sky, away from the physical world. Mythologically, the hero was born as the bearer of the Masculine, characterized by action, will, analysis, striving, and competition. Individually, boys and girls grow up, leave their parents, and establish independence in the world. Culturally, hierarchical men's societies formed as a force for the development of masculinity, of man's awareness of himself. Bravery and willpower were proofs of masculinity and ego stability. The male group with its initiation rites forged individuality and gave birth to the patriarchal culture as a whole.[15]

But as the organ of awareness, the ego functions as the center of consciousness in both males and females. Thus far, Jungian theory describes the separation of the boy from the mother, along with the development of his separate sexual identity. Preliminary developmental theory for women suggests that women separate from the maternal consciousness through developing a relationship with their animus (the inner masculine principle), which acts as a bridge to the outer world. In their book *Female Authority: Empowering Women Through Psychotherapy,* Polly Young-Eisendrath and Florence L. Wiedemann describe the stages of animus development as they relate to ego development in women.[16]

Many philosophers and psychologists consider the emergence of a separate ego to be the peak of human evolution.[17] But when the ego of Western humanity divorced itself from the world of nature to establish a separate identity, something inevitably was left behind. This separation of the ego from the world broke in two the unconscious unity of sky and earth, light and dark, mind and body, conscious and unconscious, Masculine and Feminine—casting the latter half of these opposites into the shadow. Similarly, during adolescence the ego continues to strengthen and differentiate from the unconscious. Along with an increase of consciousness and an expanding capacity for action, there is the simultaneous exclusion of the disruptive forces of the unconscious. Such separation is not wholly reprehensible. It can be seen as a necessary phase of development in which the mind needs to distance itself from the body and matter in order to look at it from above. While it is all intermingled, there can be no perspective. Such separation is only damaging when it endures beyond its necessary life span within the psychological development of an individual or a culture.

In addition to the separation from maternal consciousness, another area of differentiation is that of psychological type. Each individual (or culture) adopts either an introverted or extroverted attitude toward the world, as well as a preferred mode of functioning such as thinking, feeling, intuition, or sensation (these will be discussed in more detail in chapter 3). The undeveloped functions are cast back into the unconscious and remain in the shadow of the personality. As people begin to assume the responsibilities and burdens of the outer world, the youthful sense of unending possibilities begins to diminish, the energy wanes, the wonder fades. The ego builds structures, defends itself, and thus hems itself in.

Development of the ego allows us to deal with loss of persons and objects because we have a distance from them, but we become increasingly alienated—from nature, and from our unconscious. That is the price we pay, the more ego we develop. We lose our roots in something larger than ourselves. In the beginning, freedom is great fun, as we surge toward becoming ourselves. At midlife, however, many people give up the freedom of ego they gained in their late teens and early twenties. The sense of alienation from the whole begins to hurt and they become frightened. Something seems to have gone wrong. All the wonder, beauty, creativity, and limitless possibilities have faded. This can lead to depression, a lack of available energy. In response to this depression, there is a great tendency toward regression, to try to return to the nostalgically secure past, to revert to methods that worked in the past.

At this point, advancement of consciousness is served by reconstructing a symbolic connection with the greater energy system we have lost, building a living bridge to the nourishing Feminine we left behind. This does not mean going back to the *participation mystique*. We do not give up our ego, but enrich it through a better connection with a larger background. With heroic qualities, some individuals who take this path rescue the lost maiden from the den of the dragons—the dragons of security, status, systems of belief, religious dogmas, expectations, and norms of society—thereby releasing creativity, wonder, and new energy.

The rise and domination of the ego brings about a one-sidedness of consciousness. As the Masculine emerged as the ruler of differentiation, the Feminine retreated underground and became, in effect, the ruler of the unconscious. Like the eclipse of the moon, the Masculine eclipsed the Feminine and drove it temporarily into the shadow. While Freud regarded the unconscious as a garbage dump teeming with sexual repressions, Jung found a source of creativity and renewal in the unconscious. Early in his relationship with Freud, Jung wrote that it was his view that the objective of the depth psychological movement should be the rescue of the Feminine. In principle, the time must come when that which has gone underground must come up again.

In this context, the emergence of the Feminine can be seen as the next natural step in the evolution of consciousness. By going into eclipse, the Feminine lost prominence and power. But there are certain experiences one can have only in exile or in the shadows. Over the centuries the Feminine has gathered a different kind of power and wisdom. As the Feminine emerges again, it does so as a far more conscious force. Rather than bewail the eclipse of the Feminine, let us deal with its powers as they exist in the shadow and explore how we can assist their emergence.

Historically, science has been an important force in the evolution of consciousness. Rather than act out of superstition or take the word of the church authorities, science examines the world directly and tests assumptions. By viewing the world in terms of polarities, and reducing nature to its component parts, science has accumulated vast knowledge and power. We have now come to a point where the Masculine and Feminine are interacting, and we can look forward to their union giving rise to something greater and more important than both—something entirely new.

I believe that the broader perspective of individual psychological development and the evolution of consciousness in our culture calls us to a higher goal than simply encouraging more women to enter science. This is a joint task, involving men and women equally. As we examine the direction of development of consciousness and the psychological history of humankind, we see more clearly how we have benefited from developments of the past. We can also see our present tasks, where we need to place our attention, in order to take the next step in the process.

It is the premise of this book that, just as each individual is enriched by access to both Masculine and Feminine, so are cultural institutions such as science. Now we shall examine how science became identified with the masculine propensity to discriminate, leading to an objective, reductionistic way of studying nature.

THE BIRTH OF SCIENCE

Although the Latin *scientia* (*scire*, to learn, to know), in its broadest sense means learning or knowledge, the word *science* has come to refer to natural science, knowledge about nature. In the seventeenth century, Western science further narrowed this definition to encompass only knowledge gained through a specific procedure, the scientific method. Today, most scientists endorse the definition that science connotes a process or procedure for making inquiries about our world and for evaluating the hypotheses these inquiries generate—the scientific method.[18] Yet, there is growing recognition that science is a cultural and social product. College courses in the history of science stress that the process of science is shaped by the attitudes, priorities, methods, and beliefs of those involved in the process. In this sense, science has a decidedly masculine face.

The stereotype of the man of science is not the macho image of the Marlboro Man—rugged, muscular, clean-shaven, and strong-jawed. On the contrary, the authority of science rests on the power of the intellect, rather than on physical power. The stereotype of the scientist is a gaunt, unemotional, bespeckled man of reason with hunched shoulders, wearing a white lab coat. Until recently, a woman scientist was an oxymoron, a contradiction in terms.

Roots of the male priesthood of science can be traced to medieval Christianity.[19] Early devotees of science toiled away in ecclesiastical academies, studying the natural world in their quest for divine illumination. Fervently searching for God's truth, the monks of these ascetic mendicant orders gave rise to a science steeped in their celibate, homosocial, misogynous culture.

With the coming of Christianity, the story of Adam and Eve replaced Aristotle's cosmology, but the basic concepts of male superiority and female inferiority remained unchanged. "The Woman in us, still prosecutes a deceit, like that begun in the Garden; and our Understandings are wedded to an Eve, as fatal as the Mother of our miseries," wrote Joseph Glanvill, one of the chief promoters of the Royal Society of London.[20] As a speaker for one of the first scientific societies in Europe, he maintained that truth does not have a chance when "the Affections wear the breeches and the Female rules."[21] The ideal scientist was unemotional and detached. His tools were logic and analysis. The Royal Society formed in 1662 to provide a forum in which natural philosophers could gather to examine, discuss, and criticize new discoveries and old theories. The business of the society was "to raise a Masculine Philosophy . . . whereby the Mind of Man may be ennobled with the knowledge of Solid Truths,"[22] wrote Henry Oldenburg, the society's first secretary and editor of the first professional scientific journal. By their own definition, the Feminine had no role in the Royal Society. In fact, the English scientists disparaged their rivals in France by calling them "effeminate," and Francis Bacon attacked Aristotelian philosophy for being passive, weak, and "feminine."[23]

The burgeoning work of the seventeenth-century scientists sprang from the empirical method of science first set forth by Francis Bacon (1561–1626). The inductive process—to record the available facts, devise experiments, draw general conclusions to explain the phenomena, and then test these conclusions by further

19

experiments—became the basis of the scientific method. As a vigorous proponent for this method "to extend more widely the limits of the power and greatness of man,"[24] Bacon became extremely influential. Although the experimental method itself may be considered a neutral and objective approach to the study of nature, it emerged within a social context that determined who used the method, the type of knowledge they sought, where they looked for answers, and how they used the knowledge. To what purpose was this method used?

Bacon advocated using the new experimental philosophy to inaugurate the "truly masculine birth of time," to lead men to "Nature with all her children to bind her to your service and make her your slave . . . to conquer and subdue her, to shake her to her foundations." Whereas the medieval scientist merely sought to exert a gentle guidance over nature's course, Bacon urged researchers to use his method to discover the "secrets still locked in Nature's bosom, . . . to penetrate further, . . . to find a way into her inner chambers, . . . to storm and occupy her castles and strongholds, and extend the bounds of human empire."[25] Bacon's writings reflect the dramatic shift in the goals of science that occurred during the sixteenth and seventeenth centuries. The integrative goals of wisdom, understanding the natural order, and living in harmony with it, shifted to the goal of dominating and controlling nature.[26]

René Descartes (1596–1650) believed that all knowledge could be derived from first principles and that mathematics is the language of nature. Taking a rational, deductive, analytical approach, he maintained that all aspects of complex phenomena could be understood by reducing them to their constituent parts. Descartes formulated a sharp distinction between soul and body, mind and matter, which subsequently became a general belief. To him the material universe was nothing but a machine governed by exact mathematical laws. He regarded physics as reducible to mechanism and even considered the human body to be analogous to a machine. Nature worked according to mechanical laws, and everything material could be explained in terms of the arrangement and movement of its parts. This mechanical picture of nature guided all scientific observation and the formulation of all theories of natural phenomena until twentieth-century physics brought about radical change. The medieval view of an organic, living, and spiritual universe was replaced by that of the world as a mechanism. Removing the soul from nature gave researchers permission to dissect the machine, a license that had been previously withheld out of respect for the sacredness of the organism.

Sir Isaac Newton (1642–1727) synthesized Bacon's empirical, inductive method and Descartes's rational, deductive method into a process now known as the scientific method. Newton stressed that neither experiments without systematic interpretation nor deduction from first principles without experimental evidence can lead to a reliable theory.

By the end of the nineteenth century, the dream of the mastery of nature for the benefit of mankind seemed on the verge of realization. A comprehensible, rational view of the world was gradually emerging from the laboratories and universities. The physicist Lord Kelvin went so far as to express pity for those who would follow him and his colleagues, for they, he thought, would have nothing more to do than to measure things to the next decimal place.

Although the experimental approach liberated science from many of the assumptions and doctrines of Aristotle and those based on faith or superstition, Western science continued to deliberately veil its feminine face. From its inception, men hailed science as a purely masculine construction. The language and metaphors of science reflect the masculine ideal of objective, rational, logical, linear thinking.[27] The Feminine was subordinated, undervalued, and denigrated. One of the primary characteristics of this masculine approach is the reductionistic approach to the study of nature.

The exclusion of women from science reflects the exclusion and lack of valuation of the Feminine. The implications of this one-sidedness are far-reaching, since science not only affects our lives materially, but it forms the very concepts we have of reality. Science defines our place in the universe. It tells us who we are and what are we doing here. In this sense, science strives to answer the same questions that myth and religion seek to answer. For example, modern science tells us that the universe began with the big bang, that the Earth revolves around the sun, that space and time are relative, that DNA is the genetic material responsible for our heritage.

SCIENCE AS A MASCULINE PHILOSOPHY

Western science is far from monolithic, yet as an institution, or a way of acquiring knowledge, it has acquired certain habits or tendencies in its outlook. Because the majority of minds responsible for constructing science were male, the institution reflects masculine consciousness.

The man of science hoped that by maintaining a detached view of the world, uninfluenced by personal desires or love of his opinion, he could accurately observe the world around him. He then used logic and reason to deduce the causes of natural phenomena, to form theories and to predict outcomes. He valued objectivity; human passions disturbed the orderliness of thinking. Attention to feelings led to inefficiency. An emotional approach to scientific investigation left the scientist open to biased interpretation of the data and led others to distrust his work. The man of science trusted only what could be reliably measured and reproduced.

From a position outside of nature, the impartial researcher focused his intellect on ever smaller pieces. Only by reducing the complexity of a system, and limiting the number of variables could man hope to comprehend its workings. Linear thinking and precise quantitative analysis created a sense of order in the apparent chaos of the world swirling around him. Simplification and abstraction led to mathematical descriptions of the principles underlying the operation of nature. The man of science reasoned that, by reducing nature to a set of equations or building blocks, he could then understand the whole by adding the equations or building blocks back together.

Seeing the regular, repetitive patterns of nature brought a sense of security. Nature could be relied upon: day followed night, spring followed winter, and a solar eclipse did not mean the end of the world. Accurate prediction bespoke the success of a theory. Science structured the overwhelming confusion of the world, giving man the hope of ensuring his survival by controlling his environment. But by rejecting anything not conforming to reason, he shuts himself off from the seemingly irrational and chaotic realm of the unconscious. In contrast, science in India included "inner science" and studied alternate states of consciousness. In the West, this domain is usually relegated to the category of pseudoscience, since it deals with phenomena that are unmeasurable by our current technology.

Determining the laws of nature led to the creation of sophisticated tools and technology to manipulate the environment. Because of the power and effectiveness of the scientific method, science acquired the voice of authority in Western culture. It replaced superstition by illuminating connections between cause and effect. Today, society looks to science to create new materials and methods and prove their efficacy, safety, and reliability. Until recently, most people believed that, given time, all problems could be solved by the application of science and technology.

Science's pride in its power over nature led to hubris. The ability to control and dominate nature went to man's head. Du Pont's slogan, "better living through chemistry," envisioned a world designed and ruled by man, where he could live in wealth and comfort. The disastrous consequences of disrupting the subtle and complex balance of nature were not foreseen.

Not only did masculine values define science, thus excluding the Feminine, science also became a tool to disenfranchise women. For example, craniology, the measurement of skull and brain size, was an important scientific field during the nineteenth century when much work was done on mental abilities and the brain. Craniologists reasoned that "intelligence" was directly related to brain size and therefore began measuring the sizes of brains, or rather, of skulls. Since women, on the average, have smaller skulls than men, craniologists concluded that women were intellectually inferior to men, and consequently were less capable of thought. Scientists told women that intellectual pursuits would severely tax their sensitive nervous systems and decrease their fertility. Prejudice against women's ability to think still lingers, even now, in fields such as sociobiology.

Ian Mitroff, who interviewed forty eminent scientists studying moon rocks from the Apollo missions, observed that science continues to be one-sided to the present day. He comments on the degree to which the spirit of the moon program was under masculine domination:

> It was man, not mankind, who in body, spirit, and soul took us to the moon, who landed on the moon, who took back some of that precious moon, and finally who analyzed that moonstuff. Nowhere in all of this was the feminine principle present. . . . We must recognize the challenge to learn how to do science with feeling, how to develop a science that in its working methodology and spirit knows what feeling means.[28]

While the scientific method is a powerful tool, it is always used in a context. The seventeenth-century model of the universe as a clock composed of independent parts has evolved with the development of more sophisticated machines, so that to many, the modern model of the universe is that of a supercomputer. This concept that nature can be disassembled and reassembled under man's control represents a reductionistic style of science that is fundamental to the masculine analytical approach. While the power and value of this approach cannot be denied, it neglects the relatedness of the parts. The whole possesses more than the sum of its parts. When a cell is dismantled into its component parts, the function of the whole is lost. When a human being is dissected into an array of tissues and organs, the essence of the person is missing. Consciousness and individuality spring from the integrated functioning of the whole. The feminine principle of relatedness can give science this holistic perspective. When the Masculine is used in conjunction with the Feminine, the researcher can focus on the individual parts while simultaneously considering their relationship to the environment.

Although science has an overall masculine bias, individuals and disciplines within science exhibit a range of characteristics along the masculine/feminine continuum. The "hard," objective sciences of physics and chemistry are seen as more masculine than the "softer," more subjective discipline of psychology. Yet the fact that some scientists argue that psychology is not a true science leads the researchers in that field to become even more masculine in their approach, in order to prove the worth of psychology as a science. In seeking respect from the rest of the scientific community, these psychologists concentrate on observable, quantifiable behavior, and distance themselves from intangible subjects, such as studying creativity or dreams.

Recently, aspects of the feminine principle have been emerging in many branches of science, both in the form of new ways of seeing the world, such as that represented by the new physics, as well as in the form of greater representation of women in science. Since women are generally more conscious of the Feminine than men, their participation in the scientific enterprise can contribute to changes at a fundamental level.

THE MIDLIFE CRISIS OF SCIENCE

As I have been dealing with personal issues of midlife, I have been struck that science itself seems to be facing the problems characteristic of midlife. During the first half of my life, I established myself in the world and settled into familiar psychological patterns. When I was thirty-six, my mother died—and the approach of my own death became a reality. My hair started graying; my face felt brittle. Although financially rewarding, my job had become routine, boring, and meaningless. I felt disillusioned and disappointed in life, grieving about some vaguely felt absence. I wanted something more, but not more of the same. Midlife crisis had befallen me. To understand my unrest, I turned to Jungian psychology, a psychology that focuses on the issues we encounter in the second half of life.

Typically, our midlife realization of life's limits shatters our youthful dream of "living happily ever after." After reviewing our life, we leave behind our earlier sense of identity, our fantasy of continuous expansion and growth, our assumptions about who we are in relation to others, our hopes and expectations for an imagined future. Achievements turn hollow. A sense of loss catalyzes a fundamental shift in our alignment with life and with the world. As our mask crumbles, there is a deep restructuring of the psyche, accompanied by the release of repressed and unconscious elements of the personality. We turn inward to reevaluate our life goals and ideals. We become more concerned with finding our purpose in life. Dedication to outer success becomes modified to include a concern for depth, meaning, and spiritual values. In the second half of life, emphasis changes from the external to a conscious relationship with the inner world.

If they allow the process to unfold, extroverts begin to develop the introverted side of their nature and find nourishment in spending time alone. Thinking types begin to discover that thinking alone has its limitations. While still useful, thinking ceases to be as interesting or challenging. At midlife, thinking types begin to develop other resources: sensation, intuition, and last of all, feeling (this will be developed further in chapter 3). While some journey within to find new parts of themselves, others may break out of external structures, such as a marriage or job, to search for their missing parts in a younger mate or external adventures. At midlife, fast-track careerists may "burn out" or "mellow out" and take time to develop their feminine side; homemakers may develop their masculine side through entering the work world.

The tasks of separating from an earlier identity and confronting repressed parts of ourselves also occur during other transitional periods in life. Unique to the midlife transition, however, is the fact that we engage our contrasexual nature, whose power has been denied and evaded: a man encounters his unconscious Feminine; a woman confronts her unconscious Masculine. The quality and meaning of the subsequent struggles and deliberate negotiations are quite distinctive of the midlife transition—and critical to its outcome.

Analogously, science has spent the first half of its "life" examining the external world of nature and establishing itself in the world as a masculine philosophy. The epitome of the conscious ego, science has been the hero in our culture, adapting, expanding outward and forward. Science has become synonymous with progress. Like the hero, science takes the initiative, surprises the enemy, takes charge using aggressive strategies, and wins glory. The heroic scientist forces the secrets out of nature by conquering and subduing her.

Just as the individual realizes the approach of death at midlife, so do we now face the possible death of our planet in the form of nuclear and environmental threats such as toxic wastes, pollution, global warming, destruction of the ozone layer—all made possible by the products of science and technology. In this sense science is poised at the threshold of midlife. We are beginning to realize we may not live "happily ever after." While we once thought science could solve all our problems, some of us now recognize its limitations. Some scholars have begun to critically review the life history of science. Some scientists grieve the loss of the dream of totally controlling nature and are beginning to explore a different relationship with her. Some writers are reevaluating the ideals and goals of science, abandoning the fantasy of continuous growth as "progress." Some researchers are beginning to study "inner sciences," turning inward to probe the depths of the psyche. A few scientists are examining the ethics and consequences of our supposedly neutral science.

As we have discussed, science has relied on observing, measuring, and thinking—and rejected feeling and intuition. In our unconscious shadow, these elements have become aligned with the Feminine. It is now our task to bring them into relationship with human consciousness. The dangers threatening successful passage through the midlife transition come from the regressive tendency for the ego's natural defenses to snap the old mask and structures back into place. Restoring this former identification affords a sense of security. This defensive retrenchment and retreat into former patterns can be seen in those who still expect that simply applying more science and technology will rescue the planet.

Successful navigation of the midlife transition does not mean abandoning science and technology, but rather bringing it into a new relationship with previously repressed elements, particularly those of the Feminine. Integrating the Feminine, however, can be expected to be a slow and difficult process. Psychologically, when a part has been repressed and unused, it remains primitive and undifferentiated. It may emerge at inappropriate times and in unsophisticated ways, causing such embarrassment that it is again summarily denied expression. Patience and faith in the long-term benefits of the repressed qualities are required.

In the final stage of midlife we reintegrate the opposites and move a step closer to wholeness. We emerge with two pieces of information that have far-reaching effects on the remainder of our life: a precise knowledge of limits, and a long-range life task. These provide the core of our new identity and purpose in life. The insights gained during midlife evolve into an ethical obligation to serve the needs of the culture as well as to honor a commitment to our inner world. The optimal outcome of the midlife transition is "the creation of a reworked, more psychologically inclusive, and consequently more complex conscious sense of identity."[29] By nurturing a flow between the ego of science and the depths of the unconscious, the scientific adventure can be a journey with soul.

As we stand on the threshold of science's second half of life, we need to ask again: What are the goals of science? To predict the patterns of nature and ensure man's survival? To improve the quality of life for humanity? To control, master, and dominate nature? To live in harmony with nature? To understand the Truth? To know God? To know ourselves? To cocreate with the Creative Force? To promote the evolution of consciousness?

I believe it is time to take up the task of integrating the Feminine into the masculine philosophy of science. Throughout this book we will examine how science has been handicapped by its one-sided, masculine approach, and now must look to the Feminine, the yin, to balance itself. From the shadow, we will recover qualities that can weave together "man" and nature, that will result in a movement toward wholeness in our perception of the universe. The Feminine has always been present in science, inherent in each human being to varying degrees, but it has been veiled. Now, some women and men are discovering the value of the Feminine in themselves and their work. They find it offers a fresh view of the world, a new way of thinking. It infuses science with caring and opens up untapped resources. It gives us more flexibility and choices, and offers new meaning that can change our priorities. In revaluing the Feminine, some scientists are changing their ideas about "progress" and what makes "good science." A few go so far as to say that our survival on the Earth depends upon integrating the Feminine into science.

2

THE EMERGING VOICE OF
THE FEMININE

THE FEMININE UNDERGROUND

In 1985 I casually attended a Saturday morning lecture on alchemy given by Stephan Hoeller.[1] I expected I'd leave at the lunch break and attend to other things in my busy life. As a scientist I was curious to learn more about our history, but I could not imagine how this archaic science could be relevant to modern life. Much to my surprise, alchemy fascinated me. I returned to the lectures that afternoon, all day Sunday, and yet again Monday evening. Over the years, I continued to read and learn more about alchemy. What I had thought of as pseudoscience, a silly attempt to change lead into gold, I discovered to be a profound symbol system for transformation, a model for uniting the opposites to create something entirely new—like the new personality that emerges after midlife crisis.

The alchemical texts picture the union of the Masculine and Feminine as representative of all the other opposites: the king and queen, the sun and the moon, sulfur and salt, thinking and feeling, sensation and intuition. The symbol system of alchemy provides a useful model for what I see as the next step in the development of consciousness in individuals, in science, and in our culture. This step goes beyond simply adding women or the Feminine to science. The alchemical model calls for a breakdown of the present structures so that the opposites can unite to form something totally new—in spite of our great reluctance to give up our current organizations and institutions.

While Aristotle's writings dominated the natural sciences for over two thousand years, there existed an undercurrent of belief in the value of the Feminine, which surfaced in the writings of the Gnostics, the Cabalists, and the alchemists. Although women were not accorded equal status,[2] the alchemist's ideal of the hermaphrodite included the Feminine. The fundamental images of alchemy were coition, the conjunction of mind and matter, and the merging of male and female. The alchemists sought allegorical, if not actual, cooperation between male and female. They achieved power through "cohabiting with the elements." As early as 1284, an alchemical work, *Aurora Consurgens,* called for rescue of the Feminine both within the individual as well as within the culture. Attributed to Saint Thomas Aquinas, the *Aurora* was written in a period of history when Western humanity was beginning to feel the burden of differentiation—the pain of separation from maternal consciousness and the world of nature.

From 500 B.C. to the eighteenth century A.D., Western alchemists used physical and chemical substances as symbols relating to the internal transformation of the human being. As the elements changed in the retort, the psyche of the alchemist changed. But those who approached the alchemical processes without the right frame of mind came to grief. Many died in accidents, explosions, or from madness because they were not duly prepared, had materialistic and self-seeking motivations, or lacked the correct intentions of dedication, consecration, and unselfishness. These are such foreign notions to modern science that they sound like superstition.

The alchemists found that their internal psychological condition and the external world had a curious reciprocal relationship. Jung described this as the phenomenon of projection, where we project the contents of our unconscious onto external situations or objects (the concept of projection will be discussed in more detail in chapter 5). In this way, the world is a screen upon which we can see the interior process of our beings projected. In this alchemical laboratory, we see our transformations occur. Jung found that the psychological processes of his patients conformed to the basic principles, stages, structures, and symbols about which the old alchemists wrote.

In many ways the alchemical process is analogous to the process of scientific revolution described by Thomas Kuhn.[3] In normal science, as in normal life, we go about solving problems and adding to our fund of knowledge. As scientists refine their theories and concepts, they construct elaborate equipment and develop specialized vocabulary and skills (just as we construct particular personality structures as described in chapter 1). Consequently, scientists become invested in the theoretical framework and resist change. Increasing specialization leads to a restriction of the scientists' vision, and the science becomes more and more rigid.

Yet at some point, an anomaly crops up, like a repressed part of the personality demanding attention. This anomaly stimulates a crisis in the field and leads to the destruction of old theories. Like the first stage of the alchemical process, which entails the breakdown of old forms, it is a stage full of conflict and confusion where all the opposites begin to interact and exchange their essences. It is a time of chaos, disorientation, and depression. Yet within the darkness lies creativity. As the old structures crumble, scientists mourn their loss. After a period of lamentation, endless possibilities emerge from the darkness. All seemingly opposed and contradictory realities and points of view are seen as part of the same reality. A plethora of half-baked new theories rise up to fill the void. This leads into the second stage of alchemy, with its revelation of abundance, beauty, joy, and wonder. In this stage, magnificent diversity coexists with the one, like white light diffracted into the many colors of the rainbow.

In the final stage of alchemy, the opposites merge and fuse to create something new and superior to any of the original forms. In science, this process gives birth to a radical new theory, such as when the collision of two existing physical theories gave birth to thermodynamics. A more dramatic paradigm change occurred when lame attempts to explain away black-body radiation failed, toppling the Newtonian world of classical mechanics. Out of the darkness dawned the quantum universe.

This alchemical process of transformation occurs not just once, but continually, encompassing more and more of the opposites, moving toward greater and greater wholeness. Just as alchemy describes the transformation of scientific theories, it describes the process of artistic creation, as well as the process of personal transformation. While the analogy to midlife crisis helps us to understand the past and put our present task into perspective, alchemy helps us to understand the process of transition itself.

Jung's studies of alchemy, Taoism, the new physics, and the psyches of his clients led him to become one of the first modern male scientists to value the Feminine in equal measure to the Masculine. He perceived in alchemy a symbolic system for the psychological process of individuation. His concept of wholeness, the goal of the process of individuation, included the integration of the Masculine and Feminine. He recognized the Feminine as the source of receptivity and relatedness, and called for its integration into a Western culture that had gone too far in development of the rational, the materialistic, and the Masculine.

MY EMERGING VOICE

When I was growing up in the fifties and early sixties, the definition of success for a woman was an advantageous marriage; for a man it was a lucrative career. In order to maintain their own status, men such as my father resisted their wives' desire to work outside the home. I vividly recall how he felt shamed when my mother took a job as a secretary. A wife with leisure was a symbol of a man's success. I learned that education for boys was a necessity; for girls it was a luxury. My mother told me that if they could not afford to send both my brother and me to college, my brother would go, since he would have to earn a living. All through high school I studied and studied, maintaining a class standing between second and eleventh, in order to get a scholarship. But I found out my father did not believe in educating women: he would not contribute toward my college education and he refused to reveal his salary, so I could not apply for a scholarship. My mother helped pay my tuition because she thought I should be a teacher, a job I could fit into having a family. Although bitter at the time, I now realize my father simply reflected the beliefs of his generation. Today he is proud of my accomplishments and of the life I have made for myself.

During my twenties, I never considered myself a feminist. In the mid-seventies my first boss bristled if I exhibited any "women's lib bullshit," as he called it. The feminists I saw seemed too strident, dressing like men in gray suits and red power bow ties, using aggressive male language. They seemed to me to want to *be* men. Although I felt a core of inner rebellion when it came to discrimination, I was compliant, soft-spoken, and quiet. I certainly advocated equal rights and opportunities for both sexes. I also knew I had benefited from the women's movement by the simple fact that I could get into graduate school: I had the opportunities, doors were open for me. I wholeheartedly agreed with many of their tenets: equal pay, equal opportunity, the impact of language on our thoughts. But I was so hypnotized by nine years of scientific training that its masculine perspective was invisible to me.

Yet, I had an obscure notion that there must be a way to do science without giving up the feminine part of myself. I did not know anyone I could talk to about this, and did not quite know how to articulate this vague sense that I had. I had not yet found my own voice. Through Jungian analysis, listening to my dreams, meditation, and reading books by Carl and Emma Jung, M. Esther Harding, Irene Claremont de Castillejo, Ann Belford Ulanov, Helen Luke, and Florence L. Weidemann, I began to comprehend the Feminine at a deeper level. I hungered to exchange ideas with others who were also thinking about things like: What does it mean to be a woman, a carrier of the Feminine, at a deeper level than the roles of sweetheart, wife, or mother? Do women have to think like men and become like men in order to succeed in science? Does the feminine viewpoint have something unique to contribute to the process of science? When I mentioned to a female colleague what I was thinking about, her eyes glazed over and she said, "That's way over my head," and quickly changed the subject. When I voiced my embryonic ideas to a male colleague, I sensed he thought they were trivial and a waste of time. It did not take much to make me feel embarrassed to talk about my ideas.

While in my teens and twenties, I thrived on discussions of abstract scientific and philosophical concepts. Girl talk bored me. In my late thirties I finally became interested in what women had to say. In the past I looked to male authorities because they spoke with more confidence. As I became more knowledgeable and self-assured, I discovered that the voice of authority did not always reflect competence. I noticed that, although women did not make such a show of it, they were equally proficient. Now when I need a doctor, tax accountant, stock broker, or lawyer, I seek out capable women because I prefer the quality of our interactions.

Finally I discovered the literature about gender and science: articles and books by Evelyn Fox Keller, Ruth Bleier, Carolyn Merchant, Ruth Hubbard, Sandra Harding, and many others. Lacking a supportive community, I held conversations with them in my head, creating an imaginal community. Later, I had the opportunity to meet several of these feminist scholars at the annual meeting of the History of Science Society in Seattle. At about that time I became involved with the Women, Science, and Technology Reading and Discussion Group at the University of Washington. In addition, I had also gradually developed friendships with other people interested in Jungian psychology. In these communities, multiple viewpoints were accepted. Each individual's perspective was heard and respected. We could explore our individual truths and experiences together. There was not just one right answer. Through these books and interactions, I began to hear the voice of the Feminine in science.

WAYS OF KNOWING

In Carol Gilligan's book *In a Different Voice,*[4] and in *Women's Ways of Knowing* by Mary Field Belenky, Blythe McVicker Clinchy, Nancy Rule Goldberger, and Jill Mattuck Tarule,[5] I found studies that looked at the developmental processes of women. I finally began to find some support for my notion that science, as a system of knowledge, must also reflect women's ways of knowing in order to be complete.

Gilligan and Belenky, Clinchy, Goldberger, and Tarule challenged the concept of taking the male model as the standard. For the first time, researchers listened to women's voices rather than assuming that the female's process of psychological development must be identical to the male's. Up until then, studies had been done by and "on" males. In the midst of considerable controversy, their research laid a foundation that gave women license to find their own way by valuing the feminine approaches of interdependence, intimacy, nurturing, and contextual thought as equally valid.

So entrenched was the habit of doing research on the "male model" that it remained unquestioned until recently—not only in psychological studies, but also in basic biological research. Even in my undergraduate research on the effects of radiation on cockroaches (the college lacked the money for any more sophisticated animal facilities), I followed the standard procedure by studying exclusively male cockroaches. Presumably, the male provided a simpler model than the female—thereby avoiding interference from the cockroach's equivalent of PMS, raging hormones, and hot flashes? In an effort to reduce the number of variables in an experiment, even females as research subjects became invisible. Virtually all the experiments on how rats learn were performed on males. Federally funded studies on the effects of cholesterol on heart disease were done on 4,000 male subjects; the 15,000 subjects in the smoking study were all men; and the 22,000 doctors in the aspirin study were all male.

Gilligan's research examined women's moral development, reframing qualities regarded as women's weaknesses and showing them to be human strengths. She helped redefine ethical behavior by showing the complex equation of community, environment, and situation that go into a woman's ethical decision, compared to the hierarchical, rule-dominated male approach. Her work demonstrated that an ethic of care and responsibility may be more natural to women than an ethic of rights relying on abstract laws and universal principles.

Gilligan found women to be rooted in a sense of connection. In contrast to men's emphasis on separation and autonomy, women tend to define themselves in the context of relationships. While men tend to be exclusionary because they value distance and autonomy, women are more likely to be inclusionary because they value connection and intimacy. Whereas men resolve conflict by invoking a logical hierarchy of abstract principles, women resolve conflicts through trying to understand the conflict in the context of each person's perspective, needs, and goals. In this responsibility orientation to morality, women try to do the best possible for everyone involved. To do so, they listen intently, often suspending their own judgment in order to understand other people in their own terms. Their affinity for the world and the people in it drives their commitments and actions. For them, the moral response is the caring response. In contrast to the depersonalized language of the science of psychology, in which people are "objects" and relationships are "holding environments," Gilligan writes that: "The study of women may bring to psychology a language of love that encompasses both knowledge and feelings, a language that conveys a different way of imagining the self in relation to others."[6]

Belenky and her colleagues studied the stages of women's intellectual development—women's ways of knowing, of acquiring knowledge. Up until then, everyone assumed women learn in the same way that men do, such as described in William Perry's book, *Forms of Intellectual and Ethical Development in the College Years.*[7] Perry followed how students at Harvard (mostly male) acquired knowledge and understood themselves as knowers. After he had constructed this scheme of intellectual development, he assessed women's development and found they conformed to the patterns he observed in males.

Perry called the first stage *basic dualism,* where passive learners depend on authorities to teach them the truth. They view the world in polarities of right/wrong, good/bad, we/they, white/black. Knowledge is given, absolute, and fixed. Gradually, exposure to diversity of opinion and multiple perspectives shakes the student's faith in absolute authority and truth. In this stage of *multiplicity,* the student's opinion is as good as any other. As the teacher challenges that opinion, and insists on evidence, students enter the stage of *relativism subordinate.* Here, they take an analytical approach to evaluating knowledge. In the final stage of full *relativism,* students comprehend the relativity of truth, where the meaning of an event depends on the context of the situation and on the knower's framework for understanding the event. Within relativism, students understand that knowledge is constructed, contextual, and mutable.

Although Perry's research showed what women had in common with men, Belenky and her colleagues felt that his studies were not designed to uncover alternative routes that might be more prominent in women. Through intensive interviews, these researchers listened to what women from both formal and informal educational settings had to say in their own terms.

Several women described, in retrospect, a world of silence in which they felt deaf and dumb (in both senses of the word). From this preverbal perspective, women wordlessly obey the authorities around them—father, husband, boss—in order to survive. They have little awareness of their intellectual capabilities and see life in terms of polarities. This stage of development is consistent with the stereotype of the passive, incompetent, reactive, dependent female.

Moving out of the silence into the first stage of *received knowledge,* women listen to the voices of friends and authorities and file information away "as is," without really understanding the ideas. Thinking that everything must be "either/or," they worry that if they develop their own powers and excel, those they love will be automatically penalized. Received knowers assume that they should devote themselves to the care and empowerment of others while remaining "selfless." They are embedded in conformity and community. Received knowers look to science for absolute truth.

In the stage of *subjective knowledge,* women conceive of truth as personal, private, and subjectively known or intuited. As a first step toward greater autonomy and independence, they listen to their instincts and feelings, their "gut," and the "still small voice within." Although still holding the conviction that there are right answers, they no longer seek the source of truth in external authorities. Inward listening and watching are the predominant modes of learning. Because of this, only firsthand experience provides a reliable source of knowledge.

With the next stage of *procedural knowledge,* women excited by the power of reason engage in conscious, deliberate, systematic analysis to ferret out the truth. They use reason and procedures (such as the scientific method) to substantiate their subjective judgments or opinions. These women are aware that gut reactions or intuitions may be not always be valid, and that some truths are more true than others. Through using reliable procedures, they realize knowledge can be shared. Within this stage, Belenky, Clinchy, Goldberger, and Tarule describe two forms of procedural knowledge: *separate knowing* and *connected knowing.* Separate knowers are oriented toward impersonal rules, standards, and techniques. These women, however, reported a sense of alienation when they no longer felt any personal involvement in the pursuit of knowledge. Thinking and feeling are split and they feel fraudulent and deadened to their inner experiences and inner selves. As in received knowers, the separate knower's sense of self is embedded in external definitions and roles such as sex-role stereotypes.

Belenky and her colleagues observed that connected knowing comes more easily to many women than does separate knowing. Unlike separate knowers, connected knowers care about the objects they seek to understand. By "understanding," Belenky and her colleagues mean something akin to the French *connaître* or the Greek *gnosis,* which imply personal acquaintance with an object, and involve intimacy and equality between self and object. On the other hand, knowledge (*savoir*) implies separation from the object and mastery over it. Rather than simply emphasize the links between ideas, connected knowers also examine the relationship between knowers and the ideas they produced. Rooted in relationship, connected knowing involves feeling as well as thinking, a kind of receptive rationality. And it requires self-knowledge to use oneself as an instrument of understanding.

The final stage of *constructed knowledge* integrates thinking and feeling, objective and subjective knowing. At this stage, women find their own authentic voice. These women integrate personal knowledge that they intuitively feel is significant with knowledge they learn from others. They abandon either/or thinking and show a high tolerance for internal contradiction and ambiguity. They realize that all knowledge is constructed and that the knower is an intimate part of what is known. Rather than suppress or deny aspects of themselves in an effort to simplify their lives, they want to embrace all parts of themselves. They avoid compartmentalizing work and home, professional and personal life. They deal with both internal and external life in all its complexity. Scientific theories are now seen as educated guesswork, simplified models of a complex world—rather than absolute truth. Constructivists perceive science as a moral art, a creative construction of facts and theories directed by both the heart and mind.

Men's Forms of Intellectual and Ethical Development (Perry)	Women's Ways of Knowing (Belenky et al.)
Basic dualism: knowledge is given, absolute, fixed	Received knowledge: information filed away "as is"
Multiplicity: unsubstantiated opinions	Subjective knowledge: truth is personal, informed by emotions, gut feelings, instincts, and intuition
Relativism subordinate: analytically evaluate knowledge	Procedural knowledge: systematic analysis using reason and procedures Separate knowers Connected knowers
Full relativism: knowledge is constructed, contextual, mutable	Constructed knowledge: integrates personal and learned knowledge, thinking and feeling

In summary, Belenky, Clinchy, Goldberger, and Tarule found women's experience was characterized by such things as contextual thinking, connected knowledge, collaborative discourse, a morality organized around notions of responsibility and care, and seeing life as a web of interconnection. They argue that educators can help women develop their own authentic voices by emphasizing "connection over separation, understanding and acceptance over assessment, and collaboration over debate; if they accord respect to and allow time for the knowledge that emerges from firsthand experience; if instead of imposing their own expectations and arbitrary requirements, they encourage students to evolve their own patterns of work based on the problems they are pursuing."[8]

Through their training and life experience, the women I have known in science have become procedural knowers or constructivists. In many ways they are similar to Perry's relativists. Both men and women reach a final stage where they realize that knowledge is constructed, contextual, and mutable. But, in contrast to men's ways of knowing, the women's ways are rooted in connection. Although Perry's scheme acknowledges the importance of the knower's framework, connection and relatedness never become the dominant themes that they are in women's ways of knowing. These developmental studies substantiate Jung's definitions of the Masculine based on analysis and logic, and the Feminine based on relatedness. In the Belenky study we see the development of reason beginning at the stage of procedural knowledge. Perhaps if Perry's studies were extended to examine men after midlife, he might find some were becoming connected knowers.

When science left the Feminine behind, it was probably because the majority of the women at the time (partly due to lack of educational opportunities) fell into the categories of silent women, received knowers, or subjective knowers. Based on their experience of women and various power motives, the founders of science generalized that all women were like that and set about proving that females were, in one way or another, inferior to males. Through the centuries some women became procedural knowers through their own experiences outside of institutions. Often their procedures were not recognized as valid because their observations were made in the home rather than in the laboratory, even though the knowledge was obtained in a systematic manner. Only over the last century have large numbers of women been exposed to procedural knowledge through access to educational institutions. With formal education, women and men both learn the same procedures and language and can communicate and share knowledge.

Science is about knowledge, and the "scientific merit" of a particular experiment reflects adherence to an agreed-upon procedure. With the understanding that experimental investigation meets standards of "good science," the scientific method makes it possible to share knowledge. While the early stages of women's ways of knowing are indeed inappropriate to science, we can now see that subjective knowers mature into connected knowers and constructivists. It is important to make the distinction between subjective knowledge, where the personal truth is absolute, and connected knowledge in which the personal truths are valued and integrated into a larger whole. With this kind of discrimination, we can tease apart the stereotype of the "emotional, irrational" woman from the emerging Feminine, where feeling and thinking are not mutually exclusive. In chapter 5 we will explore how this mature type of subjectivity is relevant to science.

FEMINIST VOICES

While I do not intend to extensively review the literature, I want to give a sense of the scope and flavor of the feminist voices discussing the role of women in science.[9] Under the label of "feminist" fall various points of view, intentions, lines of thinking, and thoughtful analyses. Although all advocate women participating in science, I do not feel that they all speak with the voice of the Feminine. As cell biologist Diane Horn observes, "The training is pretty strong. You have to conform to the male patriarchal structure to get through."[10] Therefore, I have organized my survey of the literature by looking at the motivations and intentions of feminist writers: (1) to show women are capable of doing science, (2) to reveal obstacles that obstruct women from participating in science, (3) to correct the misinformation about women's biology, and (4) to critique the values and objectives of science.

The first wave of feminist literature sought to justify the existence of women in science, to document that women can be as good as men. Books such as Margaret Rossiter's *Women Scientists in America: Struggles and Strategies to 1940*[11] and *Women of Science: Righting the Record,* edited by G. Kass-Simon and Patricia Farnes,[12] show that women's brains are, indeed, large enough to do science and that education does not impair a woman's fertility (as nineteenth-century scientists had claimed). But Kass-Simon and Farnes conclude:

> For women in science to be remembered, not only must their work be thought right, but usually it must have such impact upon scientific thought that exclusion is impossible. If women scientists are wrong, or if they narrowly miss the mark, or if they propound ideas that are ultimately superseded, not only are their ideas quickly forgotten, but as often as not, the women are ostracized by their contemporaries or treated with derision.[13]

These compilations of women scientists demonstrate that nothing in a woman's physical, psychological, or intellectual nature prohibits her from doing good science. They show that many women, not just exceptional women like Marie Curie, have participated at all levels of science, from technical assistants to independent

41

investigators. We can now look to a surprising number of previously invisible women scientists throughout history, from natural philosopher Arate of Cyrene in the fifth century B.C. to fourth-century mathematician Hypatia of Alexandria, to twelfth-century Hildegard of Bingen, to twentieth-century physical chemist Rosalind Franklin, whose pivotal contribution to the structure of DNA was minimized. Girls considering science as a career now have many role models to inspire them.

As I read the stories of women scientists, I felt tremendous respect for those who struggled against such formidable opposition just to be permitted to do science. I admired their intelligence, determination, and perseverance in the face of loneliness, pain, and frustration. I felt angry at the unfairness of exclusion. At times their stories of squelched hopes, thwarted plans, and difficult compromises brought tears to my eyes. Some of the absurdities made me laugh, like when a journalist reported Lise Meitner's lecture as "Problems of Cosmetic Physics," presumably because the actual title, "Problems of Cosmic Physics," seemed so improbable for a woman.[14]

These women were so fascinated and excited by the questions of nature that they virtually dedicated their lives to science, often working without pay, often ostracized to out-of-the-way laboratories so that their presence did not fluster the men. For example, physicist Maria Goeppert Mayer worked as a "volunteer associate" in her husband's laboratory for twenty-nine years, unpaid by both Johns Hopkins University and the University of Chicago. Although the papers she wrote in her attic office during her nine years at Johns Hopkins are still quoted in chemical journals, she earned only a few hundred dollars a year—not for her scientific work, but for helping a member of the physics department with his German correspondence.

At the University of Chicago, Marie Mayer taught classes; the catalog listed her as associate professor in the physics department and a member of the Institute for Nuclear Studies. Again, the university's interpretation of their nepotism rule prevented her from receiving a salary. It is not as though she just puttered around the lab and hung out in her office while her husband did the real science in the lab. During her thirteen years at the University of

Chicago, she published papers that remain classics in the field and did the research that led to her 1963 Nobel Prize in Physics (independent of her husband). Not until 1959, when she told the university that she had accepted another position, did they offer to pay her a full professor's salary to prevent her from leaving.[15] Apparently the nepotism rules could be bent.

When women collaborated with men, stereotyping often took over. These women were dismissed as hangers-on, incapable of creative innovation, riding on the coattails of their husbands or collaborators. As recently as 1971, a well-known male physicist proclaimed at a session on women in physics at the annual meeting of the American Physical Society, "If I had been married to Pierre Curie, I would have been Marie Curie."[16] Such people are so embedded in thinking "of course women are weaker, inferior; it's obvious; it's just the way it is," they are unaware of the origins of these beliefs. I suspect most do not maliciously seek to demean women and "keep them in their place," but, more dangerously, remain unconsciously secure in their unexamined world view. Thus, harm is done in the name of good intentions, such as protecting women from the rigors of education "for their own good."

Biological determinists explained the paucity of women in science by reasoning that the inability to do science is biologically built into women. They argued that a woman's intellectual development would proceed only at great cost to reproductive development. As recently as 1982, I. I. Rabi, a Nobel laureate in physics, said that women were "unsuited to science" because they had the wrong nervous system. He said, "Women may go into science, and they will do well enough, but they will never do great science."[17] Consequently, some researchers seek to minimize the differences between men and women and argue for social equality based on biological similarity. Others recognize and value the differences between the genders and maintain that they have been shaped by history and the environment—and are not necessarily inevitable.

Some parts of the literature are pure fun to read. For instance, it is tantalizing to wonder what Albert Einstein meant when he wrote in 1901 to physicist Mileva Marič, his first wife, "How happy and proud I will be when the two of us together will have brought *our*

43

[italics added] work on the relative motion [relativity] to a victorious conclusion!" Marič's biographer, physicist Abram Joffe, claimed that he had seen the original papers on relativity that Einstein published in 1905—and they were signed Einstein-Marič. Coincidentally(?), as part of the divorce settlement, Einstein promised Marič his future Nobel Prize money, which he delivered three years later. A graduate of the Swiss Polytechnic, the MIT of Central Europe, Marič certainly had the training to be Einstein's research partner and equal. Is it possible that the theory of relativity sprang from Marič's feminine emphasis on relatedness and distrust of absolutes?[18]

Although many feminists address the question of why there are so few women in science, I marvel that there are so many, considering the educational barriers, societal norms and pressures, the glass ceiling that blocks career advancement, the poor pay (even as late as 1984, women astronomers and physicists earned a mere one-fourth of what their male peers earned),[19] the lack of recognition in professional societies,[20] the chilly reception or outright rejection by male colleagues as a "monstrosity,"[21] the exclusion from the informal communications that occur in the "good old boy networks," and the personal sacrifices required. As one of my friends said, "If you want to make money, there are a hell of a lot easier ways than doing science!"

For the privilege of doing science, women throughout the centuries have quietly coped with the inequities. Most dressed in drab clothes, and conducted themselves like nuns in white lab coats, effacing themselves so as not to distract from the serious business of science. Afraid of being excluded from participation, many women deny the high price of being successful in science. Others recognize the sacrifices to social life and personal relationships. They agonize over whether to have children, feel guilty about leaving their child in day care, are torn by the professional need to spend weeks away from their family to attend meetings and serve on study sections— and sometimes are stricken with a choice between job and marriage when the best position for each person lies in a different part of the world. Often children are emotionally sacrificed on the altar of science, paying the price of absent or separated parents. Still, the burden and blame are usually placed on the woman—rather than shouldered equally by both parents. A series of hard individual choices face women who love science.

Yet many women feel compelled to do science in order to fulfill their own sense of self. Or, like many men, they find the ultimate high in the thrill of discovery—when they see a part of nature no one else has seen before. Others want work that is interesting and intellectually challenging. Some do science for fun, like doing a jigsaw puzzle or a crossword puzzle for relaxation and intellectual stimulation. Many are driven by the passion *to know* and take pleasure in the process of knowing, the way an athlete takes pleasure in the flow of the body during a race.

Now that a number of women have entered science (particularly biology, where in 1985, 21 percent of biologists were women),[22] some feminists are using their position and knowledge to critique its biases from the inside. Evaluating science in scientific terms, biologists such as Ruth Hubbard, Marian Lowe, Anne Fausto-Sterling, Lynda Birke, and neurophysiologist Ruth Bleier show how cultural factors shape biology. As opposed to the "master molecule" dogma of molecular biology where DNA dictates a person's biology, these writers discuss how our bodies are changed and molded by exercise, diet, occupation, income level, quality of health care, environmental pollution, stress, and diseases to which we are exposed. Hubbard maintains, "Every organism constantly transforms its environment while being transformed by it, and in the case of people, the society in which we live is a major component of our environment."[23]

These critiques examine how research on females has been clouded by the desire to rationalize and maintain the social status quo, justifying women's social inferiority by pointing to her supposed biological inferiority. For example, doctors in the nineteenth century held that as a woman's intellect and brain developed, her ovaries shriveled. The feminist contextual approach to scientific measurements leads to different political and social conclusions about jobs women can do. For example, Marian Lowe discusses how the image of the frail woman has been used to keep women out of various higher-paying occupations, such as construction work or heavy industry. Recent research shows that women's lack of upper body strength comes largely from disuse, rather than exclusively from biology.[24]

Eleanor Maccoby and Carol Nagy Jacklin evaluated the enormous body of work on the psychology of sex differences. They concluded that most beliefs about sex differences were either disproved by scientific studies or still open to question. Only differences in the areas of verbal ability, spatial visualization ability, mathematical ability, and aggressive behavior had been "fairly well established."[25] In more recent work, Anne Fausto-Sterling analyzes the evidence that indicates that women are "less mathematically able and less scientifically creative than men." She concludes that these are more myths than facts of gender. Because she sees science as a social construct, Fausto-Sterling anticipates that full equality for women would profoundly alter the practice of science.[26]

A large body of literature focuses on the consequences and implications of reproductive technologies such as birth control, fertility drugs, artificial insemination, *in vitro* fertilization, and prenatal testing. Feminists examine the politics behind replacing midwives with physicians and the ways medical science pathologizes women's biology (menstruation, conception, pregnancy, and menopause are now all causes for medical intervention and control). In addition to these issues, Ruth Hubbard's book, *The Politics of Women's Biology*, explores the ideological foundations of the scientific and medical disinformation about how women's genes, hormones, and muscles function. She refers to cases where the supposed limits that biology places on women's capacity to work were used to justify a range of discriminatory practices that shunted women out of lucrative positions.

Rather than ensure that all workers have safe and healthy working conditions, employers have denied women high-paying jobs because of their potential to become pregnant. For example, when American Cyanamid decided to exclude fertile women from jobs with exposure to lead, five women were faced with the option of leaving jobs paying $225 per week plus overtime to take janitorial jobs paying $175 with no extras. These women chose sterilization rather than lose their jobs. Women have not, however, been barred from traditional (and poorly paying) women's jobs such as pottery painting in which they are also exposed to lead. When research showed cadmium and vinyl chloride to be toxic to fetuses, companies responded by excluding women from jobs involving these substances. Not until these chemicals were discovered to harm men's reproductive capabilities did industry lower the exposure levels.[27]

These cases are consistent with Geoffrey Sea's observation that standards developed by industrial hygiene researchers such as those in "health physics" (the science and practice of radiation protection) are designed to protect industry from people (that is, to prevent people from interfering with production as a result of illness or by means of lawsuits), rather than to protect employees. For instance, permissible levels of radiation contamination are now determined for the standard man with the standard mass and the standard organs. Biological criteria of damage are usually limited to the extremes of fatal cancers. Sea advocates replacing the concept of the "standard dose" with a multifaceted examination of the full variety of forms of radiation and biological responses. This would involve studying different isotopes that have different energy levels, various chemical forms of the isotopes, and the biological pathways they enter. In addition, the concept of the "standard response" could be replaced by individual assessments of risk based on age, gender, size, and metabolic type—and could recognize synergistic effects and nonfatal ailments that decrease the quality of life.[28]

Sociobiologists continue the oppressive arguments of biological determinists with such bland statements as, "The female sex is exploited, and the fundamental evolutionary basis for the exploitation is the fact that eggs are larger than sperms."[29] This has led biologists such as Ruth Bleier to critique the Wilsonian school of sociobiology[30] as "bad science," showing how it recklessly extrapolates the social behaviors of animals to human social relationships and behaviors. In her book *Science and Gender*, Bleier discusses the logical and methodological flaws made by these sociobiologists, and offers data that contradict their assumptions and conclusions.

Sociobiology sees itself as replacing psychology and sociology by attributing the whole spectrum of complex human behaviors to genetic encoding. They dismiss the contribution of culture as a thin veneer, less real than the genes that forcefully manipulate our behavior. Their two key concepts are: (1) that behaviors are programmed to maximize the ability of the body's genes to reproduce themselves; and (2) that the two sexes have different strategies for maximizing their fitness through the reproduction of the largest possible number of offspring. They reason that, since males produce millions of sperm per day, men maximize reproduction of their genes by inseminating as many women as possible. On the other hand, women have a larger investment in each offspring since females only produce one egg at a time and expend energy carrying the fetus. From this, E. O. Wilson concludes: "It pays males to be aggressive, hasty, fickle, and undiscriminating. In theory it is more profitable for females to be coy, to hold back until they can identify males with the best genes. . . . Human beings obey this biological principle faithfully."[31] In this way, Wilson and others ascribe a biological basis to marital fidelity for women and justify adultery and rape by men.

In the seventies, women began entering primatology, an area that has been important in debating and defining human nature, including female nature. Because of the stereotype of the discriminating but sexually passive female, research into female promiscuity received scant attention. Now, with careful observation of primate behavior, feminist researchers are dispelling the myth of the active, courting, promiscuous male and the passive, coy, faithful female. Their research reveals ways in which females play an active role in sexual consortship. For example, Jane Goodall describes the prodigious activity of the chimpanzee Flo, who presented herself multiple times to all the males in the vicinity during estrus.[32] In another species, a lioness may mate a hundred times per day with multiple partners over her six to seven days in estrus.[33] Although such behavior has long been observed, it was ignored in the construction of sociobiological theories.

Primatologist Sarah Blaffer Hrdy writes about her personal identification with the problems confronting a female langur. For instance, every couple of years an outside male invades a breeding unit and kills the infants, supposedly to increase his reproductive success. Hrdy identified with the victimized mothers and wondered why they put up with the new male and mated with him. Her emotional involvement led her to ask new questions, such as why females solicited males outside their troop when it resulted in infanticide.[34] Such questions prompted new observations and a reappraisal of old assumptions. Gradually, feminist primatologists produced a body of knowledge that overturned long-held tenets and unquestioned assumptions underlying theories of human evolution. Once they pointed out the male bias in animal behavior studies, there was a rush by researchers of both sexes in the United States to study female reproductive strategies. This research has far-reaching implications for social policies.

The concept of the active male and passive female colored scientists' views even at the cellular level of eggs and sperm. Up until 1980, biology texts describing fertilization emphasized the passivity of the egg waiting for the sperm to awaken it, the way the prince's kiss awakens Sleeping Beauty. Recently, Gerald and Heide Schatten, using a scanning electron microscope, discovered that the sperm does not burrow into the slumbering egg. Instead, the cell surface of the egg extends small fingerlike projections (*microvilli*) that clasp the sperm and draw it into the cell. Although the mound of *microvilli* reaching toward the sperm has been observed since 1895, it had been ignored. Through their research, the Schattens have demonstrated that the egg and sperm are mutually active partners.[35] The mammalian female reproductive tract also has been viewed as passive. Now studies reveal that the sperm must be capacitated by secretions from the female genital tract before it can fertilize the egg. Upon reaching the egg, the sperm releases enzymes that digest some of the extracellular vestments surrounding the egg—but these enzymes cannot function until they are activated by another secretion from the female reproductive tract.[36]

Such feminist critiques of cellular biology start with asking different questions—questions that do not occur to those working within the traditional framework—and being open to different interpretations of the data than those constrained by gender associations. In this way, feminists argue that their criticism carries a liberating potential for science.

In view of the long-standing prejudice against women scientists, feminists first had to establish credibility before their critiques of science could be heard as valid and not be discounted out of hand.

Before there can be a tension between the opposites that leads to transformation, there must first be an equal set of opposites.

The following chapters will take us on a voyage of discovery of a science that includes the Feminine. We will explore the values and objectives of science. We will listen to the voices of women and men scientists as they reveal to us how the Feminine can make science more creative, more productive, more relevant, and more humane. We will hear about the passion they have for their work, their delight in intimately conversing with nature, and the challenge of understanding the web of connections at every level. Stories from physicists, biologists, oceanographers, biochemists, engineers, astronomers, and atmospheric scientists show how the feminine principle that lives in *both* men and women can help restore the soul to science. We will also examine the emergence of the Feminine in chaos theory, new physics, and new biology—harbingers of a holistic science. In chapters 3–11, we will explore the applicability of relatedness, nurturing, feeling, cooperation, intuition, compassion, receptivity, and self-awareness to science. In the process we will see how the Feminine can help us discover new ways of understanding our lives and our world.

3

FEELING
Research Motivated by Love

Early in my training I absorbed the worldview that characterizes science. Unquestioning, I accepted statements such as, "emotions have no place in science," and "science is value-free." I intellectually justified my research projects and logically proceeded to solve the problems that presented themselves. While interested in my work, I felt I would lose credibility if I showed my feelings, having seen other women dismissed for being too emotional. I vividly recall a fellow biochemist scorning a neuroscientist for being "not exactly dispassionate about her work"!

Over the years, I even lost the vocabulary to describe feelings. My dreams dramatically reflected this numbness, where I had no feeling response to the horror of children being tortured. Then after being immersed in the realm of science for fifteen years, I happened to take a management course in communication. I was astonished by the fine levels of discrimination available in our language for shades of emotion and value, a language virtually unspoken in my environment.

Two articles in the journal *Science Education* from 1938 explicitly describe this emotional neutrality expected of scientists. One article calls for scientists to "deliberately renounce all emotion and desire."[1] Under the heading of "readiness to think coldly," another article advocates that scientists "be impersonal and disinterested in thinking," and "be unemotional, dispassionate, and thoroughly self-controlled in thinking."[2] Although not all practicing scientists in fact renounce their emotions, this expectation has power particularly over those who are trying to fit into the system. In addition, it is usually the unemotional face of authority that scientists show to the public.

In his study of scientists who participated in the Apollo lunar missions, Ian Mitroff asked them to evaluate pairs of adjectives relating to the ideal scientist's personality and cognitive style, such as aggressive-retiring, rigid-flexible, open-closed, vague-precise, creative-unimaginative. They affirmed the image of the ideal scientist as aggressive, hard-driving, self-serving, power-oriented, authoritarian, skeptical, diligent, and precise. But when it came to the dimension of warm-cold, 31 percent refused to even rate it because they judged it as totally irrelevant to science, saying:

> I couldn't give a damn about this; it's not interesting.
>
> This scale is completely irrelevant to science.
>
> This has nothing to do with science.
>
> If a scientist is dealing with coworkers, he should be warm. If he's dealing with a subject, he should be cold; you cannot be involved emotionally in the cold, scientific interpretation of facts.
>
> A scientist who is dispassionate and cold, allowing his whole life to be run this way, is going to lack the motivation to do science for the reason why science is done, for the benefit of the human race. I'd tend toward a warm scientist *if you can get such a person; he can run into problems if he's too warm* [italics mine].[3]

This last statement emphasizes the rarity of warmth in science, and the fine line a "warm" person must walk. Mitroff concluded that "the overwhelming judgment of the irrelevancy of this dimension can be interpreted as a suppression of affective responses or feelings."[4]

Because feeling is so foreign to science, one physics professor uses it for the shock effect it has on his students. I met Eberhard Riedel, who specializes in theoretical condensed-matter physics, at a potluck dinner of the C. G. Jung Society in Seattle. Later, over lunch at a Japanese restaurant, we chatted about his teaching methods. He said that in class he puts an equation on the board, then sits back and asks, "How does that make you *feel?*" He wants his students to "get it inside," not just copy the equation down into their notebooks. When I asked how his students responded, he laughed:

> They looked puzzled! And that is exactly what I want. I want to shock them. They may not remember the equation, but they remember something strange happened. I fully expect them to think, "What an idiot!" but I don't care. I follow it up with: "What do you feel in your bones, your belly, your fingertips? I want to get it into your intuition." The statements I make, or equations I put on the board, shouldn't be taken at face value as being right. I want them to sense a state in a more holistic way, taking into account all the assumptions, the limited applicability, and the caveats. I don't want them to come back and say, "That is the way it is."[5]

Emotions, desires, passions, attachments, and feelings have long been suspect in science as the source of bias—to some extent for good reason. Emotions can blur our perceptions. Fear and hysteria exaggerate the facts out of proportion to the situation. The desire to see what you expect to see, such as seeing a homunculus (a little man) in sperm, leads to inaccurate observations. Hanging onto a pet theory in the face of new and more insistent evidence closes the mind to new ways of looking at the world. Certainly, if we are driven by unexamined feelings, we are prone to error and delusion. While science persuades through evidence and logical arguments, propaganda and advertising exhort through the emotions. But rather than totally repress and deny feeling, let us explore how emotion and feeling can balance the one-sidedness of science and make a positive contribution by:

· Drawing attention to values and ethics
· Helping to evaluate relevance and establish priorities
· Motivating research by love of nature, rather than desire for control
· Respecting nature, rather than using nature as a commodity
· Considering the feelings of other people

JUNG'S THEORY OF PSYCHOLOGICAL TYPES

A couple years after the management course mentioned above, I became more deeply interested in Jungian psychology. My pleasure reading began to revolve around psychology, and then I entered into Jungian analysis. In addition to becoming aware of my emotions, I discovered a deeper role for feeling as a way of judging value. I also found Jung's theory of psychological types provides a useful model of the personality types drawn to science—and those missing from science. Because of the way science has been defined and practiced, it draws certain types of people. This particular way of studying nature has become self-perpetuating—so much so that the characteristics of these personality types are almost synonymous with science.[6]

Today, the Myers-Briggs Type Indicator (MBTI) is an extensively used measure of personality dispositions and preferences based on Jung's theory of types.[7] Every year over 1.5 million people take the MBTI in business organizations, educational institutions, personal and vocational counseling, and governmental agencies. Organizations use it to identify leadership styles, select teams, improve communication, and for management development and conflict resolution. Educators use the MBTI to develop teaching methods and to understand differences in learning styles and motivation. Career counselors rely on the MBTI to guide clients in choosing school majors, professions, occupations, and work settings where they can best succeed.

Jung described four modes of orientation to the world: thinking, feeling, sensation, and intuition. He visualized these forming two axes:

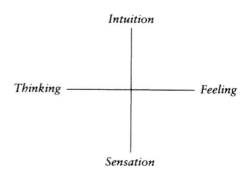

As a child, we rely primarily on one function. Some people never develop more than this one function, but as we mature, many of us develop an "auxiliary" function from the opposite axis. The thinking type, for example, may develop sensation as the second function. Later, we may develop a third function from the same axis as the auxiliary—in this case, intuition. Last to develop is the function opposite to our primary function—in our example, feeling. Of these four modes of orientation, Western science has been characterized by thinking and sensation.[8]

Sensation and intuition are ways of perceiving that supply us with information. Sensation provides us with knowledge of the external world through our senses. We apprehend the world around us by what we see, hear, smell, touch, and taste—or by extensions of our senses such as microscopes, telescopes, oscilloscopes, or Geiger counters. Intuition picks up what is in the inner world and allows us to gain knowledge or ideas in a holistic way. All the pieces of the puzzle suddenly fall into place. In a flash, all the myriad bits of data form a cohesive pattern. Intuition is open to sudden awareness, to speculation, to imagination—to all the possibilities.

In science, the sensation type is the experimentalist who gathers data and supplies us with new facts; the thinking type is the theoretician who builds models and logical systems. Most scientists rely primarily on a mixture of these two functions. (We will explore the role of sensation and intuition in science in chapter 9.) For example, a 1989 report published by the American Association for the Advancement of Science states, "Scientists believe that through the use of the intellect, and with the aid of instruments that extend the senses, people can discover patterns in all of nature."[9] They make no mention of intuition or feeling.

Both thinking and feeling provide ways of evaluating and judging. Thinking is a function of logical discrimination and impersonal choice. It draws intellectual conclusions and constructs order by reflecting on objective data. It tells us the next logical step to take. It is the function used to organize our day, to develop the detailed research plans required in grant proposals, to outline a logical progression of experiments for the next five years. Thinking is a problem-solving function. It likes to arrange things, facts, or ideas in a meaningful sequence or in a hierarchy—such as classifying all the plants and animals on the planet into a system of phylum, order, class, genus, and species. If this sounds dry, that is because thinking is dry. But it is also incredibly powerful.

The predominance of the thinking function in science has given rise to the norm of impartiality—that a scientist be concerned only with the production of new knowledge and not with the consequences of its use. For example, the thinking function is evident in the process of writing and reviewing proposals for grants. Peggy Johnson, Senior Staff Scientist at a biotechnology company, talked about the way grants are evaluated by the National Institutes of Health. As a recognized expert in her field, she serves on a study section, a peer review panel that evaluates grant proposals for scientific merit. The panel's role is to determine whether the proposal is realistic and scientifically sound—in other words, a well-designed experimental plan with achievable aims that is firmly grounded in preliminary data, accompanied by evidence that the investigator has the tools and expertise to carry out the project. The score assigned to the proposal by the study section determines, to a large extent, which research gets funded. Johnson observed that:

> The way people review these grants is not to look so much at whether the scientific idea is important, as to look at whether the actual scientific plan is realistic, where it's clear that one thing leads to the next. A premise of the whole review process is that most basic scientific research is inherently worth doing without prejudging its ultimate payoff. That premise extends to the idea of what should get funded.[10]

The thinking types' actions proceed from intellectually considered motives. Their ideal methods are governed by an abstract and depersonalized set of formal rules. They prefer to make decisions objectively, on the basis of principles and statistics. They are skilled at tasks requiring precision and organization, at seeing similarities and generalizing from them. They excel at establishing order and bringing clarity to a situation—creating models or theories from a morass of data, or ascertaining the "laws of nature." For instance, thinking types can tell us *how* to go about organizing the researchers and building the equipment to efficiently sequence the human genome.

The abstraction of the thinking function is reflected in the language used in military research, a form of techno-speak that distances the scientist from the consequences of the work. Mass murder, mangled bodies, and unspeakable human suffering are referred to as "collateral damage" from "surgically clean strikes." "Clean bombs" are nuclear fusion devices that unleash more destructive explosive power than fission devices. They are "clean" not because they kill fewer people, but because they release less radiation. In her "close encounter" with nuclear strategic analysis, psychologist Carol Cohn recounts her process of being inculcated into the thinking of defense intellectuals during a year at a university center for defense technology and arms control. As she learned their language of abstraction and technical jargon, she found herself transformed by it, entering a mode of thinking exclusively in terms of military power and technological means. She too became caught up in the pleasure of belonging to an elite group, proud of her ability to speak this "racy, sexy, snappy" new language. But she found that the language of their discourse made it impossible to ask questions of ethics or values. Although committed to staying aware of the reality behind the words, she was shocked to find she could go for days "speaking about nuclear weapons without once thinking about the people who would be incinerated by them."[11]

In an individual with a highly developed thinking function, activities dependent on feeling—aesthetic taste, artistic sense, cultivation of friends, time with family, love relationships—are usually underdeveloped. It is not that a thinking type does not have deep emotions, but rather that they prefer basing judgments on impersonal rationales. Their expertise in abstraction and generalization neglects the unique human individual and tends toward "statistical thinking." This type responds positively to words such as objective, criteria, laws, principles, categories, standards, analysis, and firmness. Since thinking types make up the majority of scientists, and the institutions of science prefer this impersonal basis of choice, scientists are often seen as remote intellectuals and science itself has developed the reputation of being "cold" and "heartless." For example, the science officers in the television series *Star Trek* are portrayed by Spock, a half-Vulcan (Vulcans are purely logical beings), and Data, an android incapable of feeling.

THE FEELING FUNCTION

Jung uses the term *feeling function* to refer to the way that we evaluate what something is *worth*. As translated from the German, it might more accurately be called the *evaluating* or *valuing function*. It is not simply emotion. As part of the human structure, emotions such as joy, excitement, anger, fear, and shame inform all the functions. The feeling function bases judgments on the personal impact something will have on the people involved. As the psychological assistant to an engineering school, Anne de Vore reports what happens in the absence of this sense of relatedness that comes from the feeling function:

> While the chairman of the chemical engineering department was on sabbatical in England, the eleven department members voted (six to five) to install a new chairman. None of the faculty informed the old chairman that he had been displaced. When he returned to the department, the faculty wives took the chairman's wife out to dinner and told her that her husband was no longer chairman. That's how he found out. These men were all thinking types and could not deal with feeling. When the dean of the school had to weed out and expel unworthy students, he didn't have the courtesy to tell them personally. He had his secretary tell them and left her to deal with the crying kids.[12]

The feeling function creates warmth where there is coldness, beauty where there is ugliness. In its extroverted form, it leads to social skills of persuasion and builds bridges between people. Feeling types tend to respond positively to words such as subjective, values, humane, good or bad, extenuating circumstances, harmony, context, and quality.

Feeling types have a strong sense of values, reacting spontaneously to people, events, and ideas. Feeling judgments have their own rationality based on a finely honed sense of good and bad, right and wrong, beautiful and ugly, and levels of importance and harmony. Such judgments depend on the context of the situation, rather than on a prescribed set of rules. As such, they can provide the basis for evaluating priorities and ethics in science, which has long laid claim to being "value-free." However, the rationale of feeling judgments frequently appears unreasonable to the thinking type because it is based on values rather than logic, and in this sense they speak a different language. When something is important to a feeling type, nothing will stand in the way of doing it. A logical rationalization or conclusion will be rejected if it feels bad or wrong. For example, the intrinsic value of the rain forest or animals facing extinction may make arguments about lower profits for the timber industry seem irrelevant.

In the process of learning to use my feeling function, I found that my body can be a sensitive instrument. In matters of aesthetics, my body tangibly responds: it feels soothed and "smooth inside" gazing at a flower and feels jarred by colors that clash. When considering the ethics of a situation, a twinge in my gut tells me "that doesn't feel right." Other people may respond with a sense of pressure or discomfort. They may feel their body hesitate, or respond with fear, alarm, pleasure, or excitement. Still others may look to their body simply for an instantaneous positive or negative feeling of attraction or repulsion. In these ways, body feelings and emotions can inform the feeling function in order to make value judgments. In conversation, I have noticed that when feeling judgments are called for, listeners will pause before responding because they need that time to "run it through their body." This time lag often leaves feeling types at a disadvantage, because thinking types usually respond rapidly from the intellect.

Physicist Eberhard Riedel says with a twinkle in his eye that he consults his body about the correctness of an equation. Although his body often says, "I don't understand this," Riedel listens to its feelings before publishing his theories:

> An equation for me is something dynamic, with many variables. You write an equation and express one variable in terms of a number of other parameters, some of which we think of as fixed or constant. In your mind you turn the knob and something happens to one of the variables that you connect with your sense of the physics it is supposed to describe. I have to take it inside and ask "Can that be right? Does that agree with my sense of the process?"[13]

By consulting his body about an equation, Riedel shows how the body can inform the feeling function in order to judge whether something is right or wrong. Gradually, we can learn to gather information from all parts of ourselves, not just our heads.

In science, feeling is the function that asks whether a project such as the Human Genome Project is worth doing—whether all the genes on the human chromosomes are *worth* sequencing—while the thinking function is interested in sequencing the human genome "because it is there" and holds to the notion of "science for the sake of science," "knowledge for the sake of knowledge." The thinking function gives us the premise that "all scientific research is inherently worth doing." On the other hand, the feeling function asks about the priority of the project and the consequences of the knowledge: What will we do with the knowledge of the human genome? Who will have access to that knowledge? Will people be discriminated against in applying for jobs or obtaining insurance if their profile contains genes for alcoholism, schizophrenia, or Alzheimer's disease? What research projects are we sacrificing in order to divert funds to sequencing the human genome? Is it necessary to sequence the entire genome, or should priority be given to selected genes? Since 95 percent of the human genome is currently called "junk," is it more important to discover the meaning of this junk or to focus on genes believed to cause particular diseases? My intention is not to attack this particular project, but simply to give an example of questions that are raised by a feeling type in the role of critic. Many scientists dislike these difficult questions about what we value because they can be showstoppers. In the wake of such a question, there is often an embarrassed silence, an evasion of eyes—and then the conversation plunges back into *how* to accomplish the task.

James Watson set an unusual precedent, however, when he became director of the Human Genome Project at the National Institutes of Health. He announced that a few percent of the project budget would be earmarked to address the ethical, legal, and social issues raised by the research. To my knowledge, this is the first time that funds have been allocated from the beginning of a project to address social consequences as the science is going forward. The National Institutes of Health–Department of Energy (NIH-DOE) Joint Working Group on Ethical, Legal and Social Issues have assembled interdisciplinary task forces and commissioned papers to define the issues and develop policies. Workshops have brought together people who ordinarily do not talk to each other: research scientists, clinicians, bioethicists, and humanities scholars. As a result of such interactions, some molecular biologists are now confronting the most probable short-term outcome of their research: due to probable lag times of years or decades between defining a genetic sequence and treatment of genetically linked disease, prevention of genetic "defects" or disease susceptibilities will be accomplished by prenatal diagnosis and abortion of "defective" fetuses. Thus the issues raised by the knowledge of the human DNA sequence are issues of value and worth that compel us to struggle with fundamental questions about the meaning and purpose of human life. For example, while our culture values physical health and intellectual abilities, it largely dismisses the value of emotional and spiritual health—equally important areas of the human condition, about which "defective" or "disabled" people may be able to teach us a great deal. Unfortunately, pregnant women and disabled people have not yet been included in the discussions sponsored by the NIH-DOE.

Using their feeling function, other scientists are beginning to question whether "doing good science" is enough. Robert Sinsheimer, former chair of the Biology Department at the California Institute of Technology, writes,

> Institutions such as Caltech and others devote much energy and effort and talent to the advancement of science. We raise funds, we provide laboratories, we train students, and so on. In doing so we apply essentially one criterion—that it be good science as science—that the work be imaginative, skillfully done, in the forefront of the field. Is that, as we approach the end of the twentieth century, enough? As social institutions, do Caltech and others have an obligation to be concerned about the likely *consequences* of the research they foster? And if so, how might they implement such a responsibility?[14]

With sorrow, he concludes that scientists may not be qualified to cope with such questions. He calls for a focus on the social context of science and on the impact of scientific knowledge on all life on the planet. Rather than operate from the notion of knowledge for the sake of knowledge, Sinsheimer feels it is imperative to select research projects that are appropriate for the social context. In a similar vein, Robert Morison, a former director of medical and natural sciences at the Rockefeller Foundation writes,

> The scientific community has led a particularly unexamined life for a surprisingly long time. . . . In all humility, it must . . . be admitted that it is impossible categorically to deny that we may have reached a point where we must abandon the faith that [in all cases] knowledge is better than ignorance. We simply lack the ability to make accurate predictions.[15]

Of the four psychological functions, feeling is probably the least present in the working life of a scientist. In fact, the feeling function has been so neglected that many claim that science is value-free. Yet a well-developed feeling function is essential in making fine discriminations in priorities, evaluating the importance of a research project, weeding out ideas, discerning aesthetic value, and assessing the ethics of a situation. Out of all the logical alternatives, the feeling function can tell a researcher which projects are most important. As discriminative functions, both thinking and feeling are rational. The feeling function is not mushy, arbitrary, or muddleheaded—but rather than relying on a set of rules, the feeling function makes judgments that are situational and contextual.

Marsha Landolt, Professor of Fisheries in the College of Ocean and Fishery Sciences at the University of Washington, illustrates how questions of context constructively enter into the daily life of a scientist. She noted that her collaborator ". . . will get very excited about the details of a new procedure or some nifty molecular tool. I'm the one who will likely say, 'Yes, but what is the question? You're right that this is spiffy, but what are we going to do with it?' There's a different approach."[16] Rather than get caught up with using a nifty tool to simply accumulate more facts, Landolt uses her feeling function to question the importance of the potential knowledge to be gained, to prioritize it relative to other questions they might ask, to choose which work to develop. Before investing her time and energy, she wants to know how the new fact might relate to the larger picture. When I asked how her colleagues responded to these questions, she said, "It varies. Usually it's a good reality check and they are usually pretty good about saying, 'Hmmm, I really didn't have a question in mind, I just wanted to learn how to do this.' "[17]

While learning to use new tools is necessary, we are more effective if we consciously undertake the task within a larger context. Modern science excels in applying technology to solve well-defined problems. But often we get caught in a loop of asking questions because we have the tools to answer them—for the fun of it, or to score another publication. While fun and communication of knowledge are wonderful parts of science, the progress and direction of science depend on the questions we ask. Our questions express our vision, lead us into the future, focus a light in the realm of the unknown. Nature is much like a Zen master who waits until the student asks the right question; only then is the student ready to hear the answer. The question is the first step toward wisdom. Our feeling function can direct us to the important questions and help us understand the essence of nature, rather than assemble a well-organized collection of facts.

Landolt adopts a similar role in representing the feeling function when she counsels struggling students. She asks them what they *like* and *want*—what they value and would love doing:

> I have a tendency to sit down with a student and say, "Hey look, have you gotten into the wrong field?" Maybe they have never thought about it, or even if they have, after accepting grant money from us for two years they might feel like a jerk if they had to say "I've decided to go to law school." But I have counseled two or three students by saying "Hey, there is more bothering you than the fact that this experiment is not working. What is it?" Maybe that's partly because I wish someone had talked to me early on and asked, "Do you really *like* this? Are you enjoying it?" I don't know what the answer would have been, because there certainly is a part of me that doesn't enjoy it, but there is another part that really thrives on it. Being able to go back in time twenty years I might have taken exactly the same path.[18]

Based on our values, feeling weeds things out by discriminating and discarding. It tells us what we want and don't want, what we like and don't like—rather than what we rationally *should* do or are *supposed* to like. Logically we can convince ourselves of anything. In the extreme, we can justify work such as Nazi medical experiments on Jews with a list of rational reasons. But something in us tells us it's wrong. Feeling protects us from harming ourselves and others. After the feeling function decides, then we can support our decision with facts and reasons.

In a climate where research funds are scarce, scientific institutions are forced to decide which research is worth funding—other than on the basis of "good science." Peggy Johnson comments on the reviewer's discomfort with making judgments of worth based on the feeling function:

> As I see it, the role of the study section [a peer review panel] was meant to be primarily a matter of ranking grants in terms of how well they are put together, without trying to put in the factor of "is this going to give more of a breakthrough or advance science more." In the current situation, that does come into it more than it was perhaps meant to, because when at least 50 percent are good grants, but you know only 15 percent of them are going to get funded, then more emphasis starts to get put on what the study section views as important science. Particularly people who have served on the study section in the past, in a different funding climate, are sometimes uncomfortable with the necessity of making those kinds of judgments because it goes against the premise that all basic research is important. It makes it a much more painful process for the members of the session.[19]

Johnson observed that reviewers many times gave higher scores to research that promised to significantly advance scientific knowledge, improve human health, or save lives. They highly valued research from an investigator with a brilliant new way of looking at something that could result in a breakthrough and lead to better understanding of an area that had been stagnant for decades. In some cases, however, they took other considerations into account. For instance, there was resistance to giving a poor score to an older researcher who had contributed to science for fifty years. In so doing, the reviewers chose to consider the value of his past work along with the merit of the grant under consideration. In other cases the reviewers are conscious of nurturing the potential contributions of promising young investigators. For all these decisions, the reviewers must reach beyond their thinking function and make judgments of value.

The feeling function also comes into science in evaluating the ethics of research. In this sense, feeling carries a moral function. In many cases, there is a risk incurred in obtaining the knowledge. Then we must evaluate: What is the price of the knowledge? Is it worth taking that risk for the sake of abstract knowledge, for the sport of science? Is the knowledge we obtain worth testing cosmetics on rabbits, isolating social chimpanzees in metal cages for a psychological study, or risking a patient's quality of life to test a new drug? Sometimes yes; sometimes maybe not. Each case must be answered individually. Both institutional review boards and peer review panels are charged with assessing the ethical parameters of a research proposal. In many areas, this pertains to grants that are clinical in nature—to assure that human subjects are treated appropriately and enter the experiment with informed consent. Peggy Johnson said:

> One of the things we're charged with doing is looking at whether anything is unethical. Usually in this study section, that would mean using patients in an inappropriate manner. For instance, one of the grants I reviewed required performing a medical procedure on very young children. This study section is approximately half M.D.'s and there was great concern expressed that it was not ethical to subject young children to this procedure to further scientific research. Those issues are then pegged so that the NIH [National Institutes of Health] officer is informed of the problem. Also, we're certainly asked to make sure that any grants using animals in studies that suggest doing anything inhumane should be brought to attention in the review.[20]

In 1991 the study sections at the NIH were required for the first time to evaluate grant proposals for whether the sex of the subjects mattered, and monitor that the appropriate mix of men and women subjects were included in the study.

Feeling types are concerned with the higher ends and values of all activities, with the distinctive worth of an individual. Rather than favoring the logical conclusion, the feeling type carefully weighs the consequences of the situation for all of those involved. Each situation calls for a unique response. The feeling type has a keen awareness of the whole network of relationships impacted by a situation.

Sometimes it takes a crisis or a disaster for people to see they've gone too far in one direction. For example, without being balanced by feeling, the thinking-type physicists of the Manhattan Project became so totally absorbed in solving the interesting problems of nuclear fission that they lost sight of the consequent human tragedy caused by their bomb. The feeling function can lead scientists to refuse to do research they feel is wrong or might be misused by a society. For example, Lise Meitner, who first calculated the tremendous energy that could be released by splitting the uranium atom, refused to take part in the Manhattan Project. Although invited to work on developing the bomb, she opposed turning her "promised land of atomic energy" into a weapon and said that she hoped the project would fail.[21] Horrified that her discovery was so rapidly turned into such a malevolent weapon, she worked toward international cooperation to prevent the destructive use of atomic energy. After Fat Man and Little Boy (as the bombs were named) vaporized human beings and reduced Nagasaki and Hiroshima to ash and rubble, Meitner commented, "Women have a great responsibility and they are obliged to try so far as they can to prevent another war."[22] I feel this is a burden we all must share.

In more recent times, many university physicists and physics graduate students signed a petition pledging not to work on Strategic Defense Initiative ("Star Wars") research. Since the 1991 federal government budgeted 52 percent of its funds for research and development to the area of defense-military (compared to 13.8 percent for health and human services, 0.6 percent to protect the environment, and 1.8 percent for agriculture),[23] these scientists risk losing funding because of their ethical stance. In deciding which research to do and which to oppose, the feeling function can help us reframe our definitions of progress and success in science. In chapter 11 we will discuss the role of the feeling function in creating a sense of social responsibility in scientists.

PASSIONATE KNOWERS

Jung observed that love and power are mutually exclusive.[24] If we are operating from a position of love, then it does not occur to us to want to control the object of our affections. On the other hand, if power dominates the relationship, love does not enter into it. When someone says they are doing something against our will "for our own good," they speak with the voice of power, not love. Feeling brings to science a style of research that is motivated by love of nature, rather than desire for control. Wonder for the beauty of nature, a passion for the work, the excitement of learning new things, the joy of seeing a pattern emerge, the ecstasy of discovery, the zest of the quest for truth, the pleasure of relationships with colleagues—these feelings can inspire and augment the logical analysis of interesting problems.

The constructivists described by Belenky and her colleagues are passionate knowers, "knowers who enter into a union with that which is to be known."[25] Such a connection to nature arises out of a feeling of devotion for the other. In contrast to the motive of prediction and control, scientists who are passionate knowers want to know out of love—like the passionate curiosity of a lover for the beloved.

In her book, *Reflections on Gender and Science*, Evelyn Fox Keller examines the biases and assumptions built into science by exploring its language and metaphors. For instance, she contrasts the Platonic knower who, guided by love, seeks to "approach and unite" with the essential nature of things, with the Baconian scientist, who equates knowledge with power and seeks domination over things. While the alchemist Paracelsus had written that "the art of medicine is rooted in the heart," and that curative remedies are discovered by "true love," Joseph Glanvill of the Royal Society warned against the "power our affections have over our so easily seducible Understanding."[26] In contrast, the Renaissance alchemists sought cooperation between male and female, and used marriage as a metaphor for the principle of harmony underlying the relation of spirit to matter, and of mind to nature. In the sixteenth century, alchemist Giambattista della Porta wrote, "The whole world is knit and bound within itself; for the world is a living creature, everywhere both male and female, and the parts of it do couple together . . . by reason of their mutual love."[27]

Leonardo da Vinci (1452–1519), Renaissance scientist and artist, also approached the study of nature with awe and reverence. His anatomical drawings combine accurate observation with an acute sense of beauty. His studies of insects and plants were inspired by a great love for everything in nature. He wrote:

67

Great love is born of great knowledge of the objects one loves. If you do not understand them you can only admire them lamely or not at all—and if you only love them on account of the good you expect from them, and not because of the sum of their qualities, then you are as the dog that wags his tail to the person who gives him a bone. Love is the daughter of knowledge, and love is deep in the same degree as the knowledge is sure—Love conquers all things.[28]

This passion to know nature as a beloved continues to motivate an underground of modern scientists in all disciplines. Although Becca Dickstein studies symbiosis (two dissimilar organisms establishing a mutually beneficial relationship, such as lichens, which are made up of algae and fungi) primarily at the level of molecular biology, she also steps back from the minutiae and speaks in awe and astonishment about the beauty of her model system. In contrast to the stereotypical detached scientist, Dickstein feels an intimate involvement with the objects of her research as she deciphers the genetic conversation between alfalfa and its nitrogen-fixing bacterial partner:

Part of my essence, part of what makes me, me, is not only the earth and the sky and the hiking and the stained glass and the colors that I love, but DNA and symbiosis and plants have in some way become an integral part of me. I always find great pleasure in thinking about symbiosis, thinking about the ways alfalfa and its microsymbiot [nitrogen-fixing bacteria] talk to each other so that they can get together to do their thing. This science is *fun* and it holds very much meaning for me.[29]

In both her work and her personal life, the beauty of nature touches her deeply. She listens intently as the alfalfa invites the bacteria to set up housekeeping together and as they negotiate the boundaries of their relationship. Dickstein reflects that symbiosis is a uniquely feminine topic of research, since it involves two different kinds of organisms finding a way to intertwine their lives so that both gain something from the relationship, a win/win situation. In chapter 8 we will explore how symbiosis provides a biological model for cooperation.

A FEELING CONNECTION TO NATURE

As I embarked on the journey of writing this book I found some women puzzled by the question "How do you feel about your work?" They were more comfortable talking about what they did than how they felt about it. Others welcomed the opportunity to reveal the hidden world of feeling they normally leave unspoken.

One evening, over tea and cookies in my friend's living room, three women cell biologists talked about their work. Their language was warm, personal—and at times passionate. When they talked about cells, they became positively rapturous. Their voices were alternately enthusiastic and spirited, then hushed, tender, and awed. At the beginning, they exchanged surreptitious glances, slightly embarrassed by the effusive feelings that emerged. Sigrid Myrdal, a senior scientist doing cancer research at Bristol-Myers Squibb, summed up the reason for their delight:

> When you look in the microscope you see life—things move before your eyes, or you stain them and they fluoresce: beautiful keratin fibers, fibronectin, actin, or cancer cells with these beautiful arms and ruffles on the edge. *They're pretty!* There's an aesthetic involved. Most men would never understand, but I knew one guy who really did understand—and I think it was probably the feminine in him.[30]

These researchers felt related to the object of their investigation, touched by the beauty of nature, rather than coldly observing nature from a superior position. Such feelings of awe and reverence do not prejudice or bias results. Instead, they create an openness to learning. Just as love of a beloved stimulates our curiosity to know everything about them, so does love of the organism (or phenomena) evoke our desire for knowledge.

Apparently, no organism can be too lowly to be worthy of love. Ingrith Deyrup-Olsen, a seventy-year-old professor of zoology at the University of Washington in Seattle, describes with delight the object of her research—slugs:

> Most people think, "Slugs—yuk!" But I think that whenever you start to study an organism, you become overwhelmed by the beauty and complexity of it. I am always *amazed* and *touched* by the way these animals solve the tremendous problems they have, which are always really basically the same as ours. I have come to have *very* strong respect and admiration for them, and I've also found it's a *wonderful* area to involve nonscientists in. The minute you begin to show them that slugs are very complicated, interesting animals with their own needs and demands, people begin to look at them with very different

eyes. I'm very moved by the slug's ingenuity and tremendous drive to continue living. I think in the end this is what makes me go on, no matter *how* frustrating the experiments happen to be at that time.[31]

Here, Deyrup-Olsen touches on many facets of relatedness: a deep love and respect for the slugs, compassion for their problems, appreciation for the complexity of the whole organism, a warm relationship with nonscientists, and an egalitarian sense that we can learn something about ourselves by studying the way slugs meet their needs.

While most accounts of women scientists describe their scientific accomplishments, these accounts rarely explore the woman's beliefs, values, attitudes, and approach toward science. An exception is Evelyn Fox Keller's book *A Feeling for the Organism: The Life and Work of Barbara McClintock*,[32] which describes how McClintock's work embodied values of the Feminine. As a geneticist, McClintock approached her object of study with reverence and humility. Rather than separate herself emotionally from her objects of study, she became intimately involved with her corn plants. In describing her work, her vocabulary is one of affection, kinship, and empathy, rather than that of battles, struggle, or a sense of opposition. She says, for example:

> No two plants are exactly alike. They're all different, and as a consequence, you have to know that difference. I start with the seedling, and I don't want to leave it. I don't feel I really know the story if I don't watch the plant all the way along. So I know every plant in the field. I know them intimately, and I find it a great pleasure to know them.[33]

For McClintock, science is not based on a division between subject and object, but rather on attentiveness as a form of love. While many other geneticists relied on statistics and probabilities, McClintock wanted to know each individual. Far from prejudicing or impeding her work, McClintock's "feeling for the organism" brought her closer to the chromosomes she studied and enhanced her power as a scientist:

> I found that the more I worked with them, the bigger and bigger [the chromosomes] got, and when I was really working with them I wasn't outside, I was down there. I was part of the system. I was right down

there with them, and everything got big. I even was able to see the internal parts of the chromosomes—actually everything was there. It surprised me because I actually felt as if I was right down there and these were my friends. . . . As you look at these things, they become part of you. And you forget yourself.[34]

McClintock also exemplifies the feeling function's concern for consequences. When she reflects on ecological tragedies such as Love Canal, she takes scientists to task for just accumulating data without taking the time to look, and applying technology before thinking about its implications:

We've been spoiling the environment just dreadfully and thinking we were fine, because we were using the techniques of science. . . . We were making assumptions we had no right to make. . . . We're not thinking it through, just spewing it out. . . . Technology is fine, but the scientists and engineers only partially think through their problems. They solve certain aspects, but not the total, and as a consequence it is slapping us back in the face very hard.[35]

Always mindful of the hidden complexity that lurks in even the most straightforward system, McClintock cautions against making assumptions about how the whole system works based only on knowledge of some of the parts. She counsels that we approach nature with humility.

In writing about nature, scientists are expected to take the perspective of the detached observer. Written in the passive voice, free of feeling, the research paper strives to communicate nonemotional concepts. This detachment stems from the early days of the Royal Society of London. Henry Oldenburg, the first secretary of the Royal Society wrote, "French naturalists are more discursive than active or experimental. In the meantime the Italian proverb is true: *Le parole sono femmine, i fatti maschii* (Words are feminine, facts—or, rather, deeds—are masculine)." [36] Consequently, scientists pruned their speech of all unnecessary eloquence. Joseph Glanvill, a champion of the Royal Society, wrote that he is "more gratified with manly sense, flowing in a natural and unaffected Eloquence, than in the music and curiosity of fine Metaphors and dancing periods."[37]

However, a few scientists such as Rachel Carson have dared to infuse their writing with wonder. Carson could not separate the beauty from nature. She saw beauty as integral to science. When she accepted the National Book Award for *The Sea Around Us* in 1952, she said:

> The aim of science is to discover and illuminate the truth. And that, I take it, is the aim of literature, whether biography or history or fiction; it seems to me, then, that there can be no separate literature of science. . . . If there is poetry in my book about the sea, it is not because I deliberately put it there, but because no one could write truthfully about the sea and leave out the poetry.[38]

Her later book, *Silent Spring*, called into question the attitude of industrial society toward the natural world. In order to discredit this work on the environmental impact of pesticides, critics called Carson a "bird lover," a "cat lover," and a "fish lover." The fact that these terms were intended to be pejorative shows how risky it is to reveal feelings in the world of science. But her feeling function pushed her to see the consequences of the widespread use of pesticides; she cared enough about the creatures of the earth to meticulously document those consequences.

This feeling of connection to nature does not apply only to biologists and their "feeling for the organism." It applies equally to physicists such as Eberhard Riedel, meteorologists, engineers—all scientists. As a child, one woman engineer's dream was to have a job where she could get up in the morning and say, "Oh boy, I'm going to do something *fun!*" rather than, "Oh gee, I've got to do something I hate just to bring home a paycheck." She has achieved her ambition and now rhapsodizes about the uniqueness of every structure, even though all of them are supposedly built on the same plans. Her love of structures enlivens her work. She said enthusiastically, "They're just plain *beautiful creations!* I'm sitting here looking out the window at a structure right now, and it's just a beautiful, graceful, flowing thing. From a purely aesthetic point of view, I just *love* it!"

NEW PERSPECTIVES

A feeling connection to nature prevents us from treating the "products" of the earth as mere commodities to be used and discarded. Over the years I've often been appalled by some of the products that science has produced as a result of its separation from nature. For example, Weyerhaeuser Company markets a product called Interiorized™ plants—real plants, from baby's breath to sixteen-foot oak trees, that are embalmed by a chemical process that preserves them in a "lifelike state." No sunlight, watering, feeding, pruning, pest control, or repotting are required by the plants, resulting in a sterile "no muss, no fuss" commodity—without life. Another company offers a service to freeze-dry family pets so the owners can put them by the fireplace and pretend their pet never died. In fact, parts of nature's gene pool can now be patented. The U.S. Patent and Trademark office accepts patent applications for life forms. It defines genetically engineered organisms as a "manufacture or composition of matter."

In *Walking with the Great Apes*, Sy Montgomery illustrates how Jane Goodall, Dian Fossey, and Biruté Galdikas focused on differences between individuals (a hallmark of the feeling function). At the time they began their work, they were pressured to talk about *the* adult male, *the* adult female, and to refer to animals by number. Rather than discover individual motives for differing responses, ethologists theorized about the underlying mechanisms of universal behaviors. Up until the time Jane Goodall went to the Gombe Reserve, no one had been able to study chimpanzees in the wild. Trying to think like a chimpanzee, and imagining what it was like to be one, opened her up to ask new questions about individual relationships and behaviors. More recently, her love for the chimpanzees led Goodall to defend them in the wild as well as work to improve the quality of their lives in zoos and laboratories.

Jane Goodall's success grew out of her love and rapport with the chimpanzees. She learned their language of pant-hoots, grunts, barks, and facial expressions, and tried to understand their lives and relationships from their point of view. Like McClintock and her corn plants, Goodall knew each of the chimpanzees individually. She gave each of them names (Flo, Flint, David Graybeard, Fifi, Mr. Worzle), knew their families, and followed them throughout their lives. Out of love for the animals she observed, Goodall negotiated the treacherous slopes of the seemingly impenetrable forests, endured malaria, tsetse flies, and ill health. Through her patience and dedication, she was the first to observe chimpanzees fashioning tools (previously thought to be a defining characteristic of humans). Instead of descriptions of Flo frolicking with her baby Flint, Goodall's graduate advisor at Cambridge pushed for measurement and numerical analysis. When she submitted her first scientific paper, the editor returned it, insisting she number the chimpanzees instead of naming them. The editor also deleted references to chimpanzees as "he" or "she" and replaced them with "it." Fortunately, when Goodall refused to make the changes, they published it anyway. This insistence on the particular, on individual differences, comes from the feeling function, which resists "statistical thinking."

The male primate researcher Kawai Masao uses the Japanese word *kyokan* (which translates as "feel-one") to mean "becoming fused with the monkeys' lives where, through an intuitive channel, feelings are mutually exchanged."[39] Rather than just assemble tables of data and statistical information on animal behavior, this approach calls for scientists to know the individual, to travel with the animal, to see life through the animal's eyes, to allow themselves to become transformed by sharing the animal's life—to see what ordinary people cannot normally see. The work of scientists such as McClintock and Goodall indicates that the ability to identify with the object of study, to feel kinship and attachment, yields a deeper knowledge. *Kyokan* or the "feeling for the organism" does not replace the information gained by a statistical approach. Rather it complements it by using the feeling function to understand the importance and uniqueness of the individual. Just as writers speak most universally when they reveal their most personal feelings, so can scientists discover the world in a single "grain of sand." In this sense, feeling is necessary to discover truth and produce the best science.

When we think of ourselves as sons and daughters of nature, and as cocreators with nature, we think twice before abusing her. The astronaut Taylor Wang comments about his experience of seeing the beauty of the Earth from space:

> They say if you have experiments to run, stay away from the window. For me, preoccupied with the Drop Dynamics Module, it wasn't until the last day of our flight that I even had a chance to look out. But when I did, I was truly overwhelmed. A Chinese tale tells of some men sent to

harm a young girl who, upon seeing her beauty, become her protectors rather than her violators. That's how I felt seeing the Earth for the first time. "I could not help but love and cherish her."[40]

A feeling of connection to nature engenders an attitude of love and respect. In the way that Goodall's love for chimpanzees compelled her to protect them, so can love for the Earth make us better stewards. With a mind-set of love and respect, research becomes a dialogue with nature rather than "putting her on the rack and extracting her secrets from her," as Francis Bacon described the virile power of science to penetrate and subdue nature.

WARMTH IN SCIENCE

While the Apollo moon scientists bristled at the idea that warmth had any relevance to their work, science is a highly social activity. It depends on people collaborating and sharing their work. The quality of interpersonal interactions and people's feelings impact the quality of work they do, although many would like to deny it and restrict their feelings to their personal life at home. The hallmark of being professional is not to show your feelings, to be competent and factual, assured and confident. This generally means not to show feelings of vulnerability. Yet in a warm environment we can allow ourselves to be more fully human. We can explore more freely without worrying about appearing foolish.

Atmospheric scientist Kristina Katsaros sometimes goes on cruises where she may be the only woman. On one such trip, one of the crew made a bad mistake, destroying the opportunity to obtain a last set of data of the measurement program. Afterward this usually cheerful man walked around the ship with a shadow over his face. Kristina showed concern for his downhearted feelings and assured him that no one held the mistake against him personally. This gave him a chance to express his sorrow at being the cause of such a big setback. Later, the chief scientist expressed his appreciation for her smoothing things over. One of her colleagues said, "You're really very good for this experiment. Usually at times like that tempers flare. Your presence makes a difference—I've noticed it many times before." Although attending to people's feelings is not strictly part of the work, it does promote better communication, collaboration, and cooperation. Several of her male colleagues have said they like having women in the field because "people behave a little more civilized."[41]

This may seem trivial, but the feeling tone of an environment can enhance or alleviate stress. By attending to feelings and creating a warm environment where people feel comfortable and appreciated, more energy can go into creative activities and less into defensive maneuvers. To be continually surrounded by tension is draining. By attending to the shame felt by her colleague, Katsaros made it possible for his feelings to heal rather than fester. While he was constantly on "red alert," expecting an assault of blame, he diverted energy from productive work and funneled it into maintaining defensive shields. Yes, attending to feeling seems like a nuisance; it takes time and energy to deal with feelings; it is uncomfortable—but unless we do, the work often becomes stalled.

Biophysicist Cynthia Haggerty found that her warmth and enthusiasm engendered a pervasive feeling of excitement and creativity in her lab:

> I loved what I did and that becomes infectious. If I saw something with the electron microscope that was exciting and new, I'd call the lab technicians to come and look at what I'd found and ask, "What do you think of this?" One person found a section of fish egg with the micropyle where a sperm was headed into the egg. It was a fantastic picture! . . . We generated an infectious excitement. Some people in my lab became so fired up that they went off to graduate school, and then I'd have to train new people. But I had no trouble with making those shifts, because it was good for them and right for them.[42]

Haggerty's love of her work provided a powerful motivating force, both for herself as well as those working with her. Because of this love, people in her lab sometimes worked until midnight or over the weekends, for the sheer joy of sharing their discoveries with their coworkers. At the same time, they were enormously productive in terms of publications. This underlying dynamic of love created a lab full of fun and laughter, as well as hard work.

In science, as in life, we use thinking for concrete ends: to plan, to plot, to calculate, to figure things out. Feeling can broaden and uplift thinking. But alone, feeling can be a chain around our necks, tying us down with attachments and connections, paralyzing us with the complexity of cascading consequences. When thinking and feeling are alchemically conjoined, each loses its limiting quality: feeling loosens the concrete narrowness of thinking; at the same time, thinking loosens the chains of attachments imposed by feeling. If properly combined, we can respond in a creative new way that is more than simply the flexibility to choose one mode or the other as appropriate to the situation. Once united, both thinking and feeling lose their sting and are transformed. Alchemically, their union gives us the Elixir of Life.

While the sword of the intellect has played a powerful role in the development of modern science, now let us turn to see what the cup of receptivity can offer us.

―4―
RECEPTIVITY
Listening to Nature

One of the archetypal qualities of the Feminine is receptivity, symbolized by various types of receptacles, no doubt derived from the womb as a receptacle open to impregnation. Since the time of Egyptian hieroglyphs, vessels have symbolized receptacles in which the intermingling of forces takes place. The Celtic cauldron held the forces of transmutation and germination. The cup and the Christian chalice have long been symbols of containment. The alembic contained the transforming reactions of the alchemists. Jungian psychologists speak of the analyst as providing a "container" for the therapeutic process. Within the individual, the unconscious holds the swirling possibilities and potentials of our internal being, all the undifferentiated opposites. As these symbols of receptivity indicate, marvelous processes take place within the seemingly passive vessel.

The receptivity of the Feminine brings to science an openness to listening to nature and responding as in a conversation or as a cocreator with nature. This is quite a different approach than seventeenth-century chemist Robert Boyle's attitude toward God's "great pregnant automaton" as he called nature. Boyle wrote that there can be no greater male triumph than "to know the ways of captivating Nature, and making her subserve our purposes."[1]

"DOING" SCIENCE

In a culture where we brag about "working hard and playing hard," we place little value on receptivity. At an early age, we learn "it is better to give than to receive." Giving feels more powerful, more potent than receiving. Our culture values what is visible, and activity is visible. The fruits of receptivity are not, however, immediately apparent.

On the scientific fast track, many scientists feel they cannot compete if they slow down enough to be receptive. Being receptive looks like they are not doing anything. Staring out the window is equated with goofing off. Ours is a "doing" culture. In science this means doing experiments, writing grants and papers, reading the literature, teaching, administering, managing, attending committee meetings and scientific conferences. There is unrelenting pressure to produce, and to produce something visible: data, papers. Fast. Many scientists feel self-deprecatory if they're not working sixty hours a week and rushing madly to do their next experiment.

Modern science is basically an impatient field. Few scientists have the patience for the questions that take a long time and require a lot of thinking, listening, and not just doing. Patience involves long-suffering, submitting to a process. Watching a slug for hours on end does not seem very efficient or productive. Rather than allow nature to unfold before our eyes, we prod and push, dissect and extract.

Of course this varies from field to field, from one individual to another. Of course scientists stop to think about their data as they are writing papers and grant proposals. Receptivity certainly exists in science as researchers look at data, read and absorb the work of others, and listen to colleagues. But instead of being an equal partner to aggressive activity, receptivity usually takes a back seat. Rather than relax and court inspiration, or take long walks while data simmer within the cauldron of the unconscious, there is a compulsion to rush out into the lab and do the next experiment. It seems outrageous even to suggest that perhaps science would be further along if people did fewer experiments and spent more time on the beach reflecting on their work.

Underlying all this activity churn a number of fears: If you do not look busy, you may not be seen as serious, dedicated, or committed—and you may be given more teaching assignments or committee work. Doing the next experiment shows you're really "on top of it" and in control, while stepping back and reflecting

79

seems an admission of uncertainty—that you may not know what is going on or what to do next. If nothing happens, time feels wasted. There is a fear that if you stand still in science, you'll get left behind. Receptivity depends on an implicit trust that there is something of value to be received.

While receptivity requires a halt to busyness and activity, it is not passive. It takes patience, alert awareness, openness, and responsiveness. As implied in the symbolism of the vessel, the receiver contains something while a process of combination and recombination occurs. From the reactions, transmutations, and transformations within the vessel, something entirely new emerges. Implicit in receptivity is discrimination—the ability to say no, to reject, to refuse to receive—using the feeling and thinking functions to judge what to receive. Being receptive does not mean being a garbage pail that accepts everything. In science, receptivity is important in the forms of observation, reflecting on data, and listening to nature. It also means receiving others and being received by them.

Rather than being in control and manipulating things, receptivity relies on observing, letting it happen, allowing something to unfold in its own time. It may seem vulnerable, powerless, and unheroic. Receptivity cannot be hurried, forced, or pushed. It involves openness and waiting. A woman engineer describes her way of designing a structure as a receptive process, using earthy, organic language that is distinctly different from the authoritarian language of control: "My experience of it is more of a letting it happen, and letting it grow, rather than forcing it. There's something about a project that wants to come into being. You can try too hard to form it."

The activity of workaholic scientists can be a way of avoiding feelings. By focusing on solving the important intellectual puzzles of nature, we can ignore our personal and internal lives. In stopping to listen and reflect on the science, long-repressed personal material may come up first and demand attention. Feelings of anger, dissatisfaction, or a sense of meaninglessness may emerge. This experience can be frightening and so is quickly dismissed as a waste of time, an inconvenient distraction irrelevant to the science. But in addition to receiving nature in the form of openly perceiving data and information, we must also receive ourselves.

Now let's examine the strength and power inherent in the feminine quality of receptivity and explore how it applies to the daily practice of science.

A VOYAGE OF DISCOVERY

An attitude of openness leads the scientist on a voyage of discovery. Without preconceived ideas, scientists find strangeness in nature they could not have anticipated in their wildest imagination. For those with eyes to see, new realities unfold. Unfortunately, because some strange observations or ideas do not fit with current beliefs, they often languish for decades or centuries on the fringes of science. For example, Western astronomers literally could not see changes in the heavens because they believed the heavens to be immutable. No one recorded the appearance of new stars until after Copernicus opened people's minds to the possibility of celestial change. It was not only Galileo's invention of the telescope that allowed astronomers to "see" more. At a much earlier date the Chinese, whose cosmological beliefs did not preclude celestial change, had recorded changes in the heavens such as sunspots and the appearance of new stars.[2]

Primarily observational sciences such as astronomy rely more on receptivity than the established experimental sciences. Unlike manipulating or dissecting laboratory rats, it's a bit harder to push and prod the universe. Astronomer Paula Szkody was drawn to study variable stars because she liked the concept that you never know what it is you're going to find:

> Variable stars are always doing some things different. They're not predictable. The ones I work on are probably the most unpredictable of the variable stars. They don't do regular things. I'm drawn to that aura of the unknown, the uncertain, and there is always something new to be found. Once you have the observations, you have to try to explain what is happening.[3]

This sense of openness can lead to new insights and new ways of looking at the world. The logical and analytical conscious structures of the mind have great value, but they are also limited. By giving up our conscious purposes, letting the seed of an idea fall into the fertile unconscious, and trusting in the process of growth, we can bring forth great creative fruits. Theoretical solid-state physicist Eberhard Riedel believes that receptivity is as valuable to the scientist as to the artist. "Without receptivity, without this kind of softness, how do you allow the new thing to come into you? This question of creativity is close to me and I want to approach it in a kind and tender way."[4]

While the artist and the scientist are often viewed as anchoring opposite ends of a spectrum, the most successful scientists look at their work as an expression of creativity. They create new theories and new understanding about the world that give rise to new technologies, which in turn create a new world for us to live in. New concepts of reality that emerge from science influence our philosophy and the way we lead our daily lives. Just as scientists need this "softness" to apprehend something new, so must the institutions of science (funding agencies, journal editors, and scientists involved in the peer review process) receive the new offerings of individual scientists. Unfortunately, the conservative and skeptical institutions of science often subject many new theories to a hazing or dismissal rather than a respectful reception.

Developmental biologist Aimee Bakken found her receptive approach ran counter to the traditional approach in her field:

> When I was doing my dissertation, my advisor kept wanting me to make predictions about the way the data should come out. I didn't want to do that. I wanted to wait and see. I didn't want to put any preconceived conditions on it. . . . It's not a matter of trying to force specific answers on it, but the answers won't reveal themselves unless you believe in it and keep working at it.[5]

Even Bakken's language of allowing the answers to "reveal themselves" is different than the image of "hammering away" at a problem that I constantly hear in science. Instead, Bakken prepares the way by focusing her attention on the problem and invites the answers by listening. Then suddenly everything shifts and a part of nature reveals itself. In contrast to the Baconian language of conflict, Bakken likens her approach to science to "opening up a flower to see what is inside," where the process of opening implies a gentleness and sensitivity to the beauty of the flower.

The scientific method, the foundation of science, consists of forming a hypothesis and then checking it out to see if it has predictive value. Biophysicist Cynthia Haggerty's work took on a different quality as she studied the pathologies of fish under the electron microscope. Haggerty found her work succeeded much better when she had a general overall question and then set up an open-ended testing system. When she sat down at the electron microscope, she asked, "What is here for me to see? What is here for me to know? What does this material *tell* me about the life process and about whatever pathology is going on?" This attitude of receptivity is essentially a devotional attitude, one of coming to the material with reverence and asking questions with humility and respect. In return, Haggerty spoke of being "given something":

> Then I was free to see what indeed was there, rather than to have a presupposed idea of what I wanted to find. So I would go through and do all the photography and stack up all this material revealing the "what is," then begin to sort it, characterize it and say, "It looks like this is happening. Can we test it out to see if indeed it's true?" Only then did we do the tight experiment to a get yes/no answer. A lot of the work I did was forging a new field. There was a lot of material that was not testing out a hypothesis that had already been formed, but doing the observational work and creating a new area of work. I loved the work! It felt very creative to me.[6]

After the initial "Oh, wow!" it takes experience and knowledge to recognize that a golden egg has fallen into your lap. And then it takes a lot of logical steps, analysis, and synthesis to put it all together. For Haggerty, this meant fitting all she saw back into the living system and figuring out what it said about the function of the whole organism. She observes what works or fails, and wonders why, incubating the observations within the container of herself. In areas where her intuition functions optimally, her mind has been prepared by years of reading, observation, and experience. Then when she receives new observations, she finds that ideas form in an organic sort of way, which includes logical analysis but is not driven by logic. Only at that point does she begin to ask: is it this or that? She feels researchers cannot work effectively in any new area of knowledge—and science is about new knowledge—with pre-conceived ideas.

Often we proceed too rapidly from the observational phase to drawing up hypotheses, making predictions and doing the "tight experiments." It takes a certain faith in the value of the process to suspend our conclusions and allow other possibilities to bubble up from the unconscious. Cancer researcher Sigrid Myrdal describes her approach as "not having to figure it out right away, but just sort of listening—just listening, without knowing how you're listening."[7]

Unfortunately, our system of funding science depends on detailed three-to-five-year experimental plans based on a series of predictions of how the experiments will turn out. This makes science more of a planned tour than a voyage of discovery. Researchers who do not get to their projected destination often feel like failures. Under the freedom of a more relaxed, open-ended approach, even "negative" results tell us something about nature. In industry, the demands of the market place, of making immediate profits, usually preclude "blue-sky" research. Only large companies such as AT&T could afford large research institutes devoted to more open-ended, long-term exploration. Now, as a result of the diversification, Bell Labs is also becoming more market-driven, with the success of the experiment ultimately defined by the bottom line. As one scientist said, after moving from academia to a biotechnology company, "Negative results aren't as much fun as they used to be." Focusing on rapidly developing products leaves little time for receptivity. Science on the fast track greatly diminishes our flexibility and ability to respond to nature.

LISTENING TO THE "NOISE" OF NATURE

Often the data taken from nature go into instruments and onto computers, unseen by human eyes. With computer modeling, the noise of nature can be literally programmed out. For example, James Lovelock writes of how "big science" hampered the discovery of the ozone hole. Huge sums of money were spent on satellite, balloon, and aircraft measurements and expensive computer models of the stratosphere that vastly improved our understanding of the atmosphere. Regrettably, the computer modelers were so sure they knew all that mattered about the stratosphere, they programmed the satellite instruments to reject data that were substantially different from the model predictions. The instruments detected the ozone hole, but those in charge of the experiment ignored it, saying in effect, "Don't bother us with facts. Our model knows best." It took a lone pair of observers using an old-fashioned and inexpensive instrument during an expedition in the Antarctic to "see" the ozone hole.[8]

When researchers obtain the results they expect, they are more likely to believe the data and less likely to repeat the experiment than when they get aberrant results. It becomes particularly hard to keep an open mind after things have been going in a certain direction and only one more experiment is needed to say that "yes, this the final answer." When the last piece does not fit into the puzzle, it is especially easy to ignore it. If there is a curve and a couple of points fall off the curve, it is tempting to figure they must have been bad data points—which is often the case. But, in fact, sometimes those data points are trying to say something significant in this game of charades we play with nature.

How do scientists respond when they get unexpected results— data that does not fit? Careful and rigorous scientists do the experiments again to find out if the aberration is "real." If it is, they set aside their preconceived notions and start to wonder what it implies. Good scientists constantly struggle to keep an open mind so that they do not miss a clue. After checking the instrumentation, double-checking for silly mistakes, making sure they did the experiment right, and replicating the experiment . . . the best scientists get cheerful when they get results they did not expect. Aberrant results lead them in a new direction and that is fun. Instead of dismissing contrary data as an aberration, an exception, or a contaminant, zoologist Ingrith Deyrup-Olsen laughed with delight at the prospect of obtaining data that contradicted her theories or predictions. When the slugs she studies present her with unexpected results, she says:

That's the best thing that can *possibly* happen, because that shows you that what you were thinking was wrong or too simple, and the animal is telling you that isn't the way it works at all. And I *must* say I find that very stimulating for students too, that when they come and they've got

85

a result they didn't expect and they ask what did they do that was wrong, and the answer is that whatever the animal does is right. It's *always* a kind of an epiphany for them, as it is for myself too. It's just one of those wonderful things that you learn that you don't know and the animal or organism does know.[9]

Rather than impose an answer on an experiment, and view data that deviates from that answer as a mistake to be erased, the receptive approach of the Feminine is to "listen to the material." Then science becomes a dialogue with nature rather than an inquisition. Cancer researcher Sigrid Myrdal describes how she listens to her experiments:

> There's the question of how you react when your data do not turn out the way you want them to. One possibility is to think "Oh no, something went wrong, my experiment failed," or "Did I ask the question wrong?" and put the data in the drawer. I think the feminine approach is to ask "What's this trying to tell me?" and consider that nature may be more interesting and complicated than I expected, but therefore probably a bit more elegant. By actually having to deal with the data, I've come to totally different interpretations. If something turns out quite screwy, I give it a chance. It's possible that it's more feminine to give something a chance.[10]

Willingness to sacrifice preconceived ideas and pet theories requires humility. Some people like to be right all the time, especially if they are invested in being an authority. They may become so identified with their ideas about how nature works that it is hard to admit that the idea is wrong because it means *they* were wrong. Too much ego involvement locks researchers into ideas, and they feel humiliated if proven wrong—even though the process of science requires trying to disprove theories. Sylvia Pollack says, "It's hard to remember to 'listen to the material' because none of us would be in science without having a pretty healthy ego."[11]

Contrary data have forced vision scientist Davida Teller to rethink her theoretical structuring, her reasons for having made the prediction in the first place. For her, the purpose of science is to develop accurate descriptions of natural phenomena, and to develop theories that cannot be rejected in the future.[12] In order to do this, she believes that the ability to listen to what your data are telling you is crucially important. When it comes to listening to nature, Teller says, "If your deepest goal is to find out how the universe really is, then if the data do not come out in agreement with your predictions, it's OK, because you wanted to find out what was true."[13]

When we explore out of curiosity, without preconceptions, the marvels and ingenuity of nature astonish us.

SHUNNED THEORIES

It always takes more energy and more proof to convince people to change their minds than it does for the first plausible theory to achieve acceptance. I know from my experience in industry that people demand more data and proof to substantiate changing a specification or a manufacturing procedure than they require to adopt an original process. Marketing people know that the first product in a market niche reaps the advantage of customer inertia. Customers become familiar with the product and need substantial persuasion to consider an alternative. Busy with other things, they do not take time to evaluate a new device if the old one serves most of their needs. Similarly, once a theory explains most of a phenomena and gives some predictability, many scientists ignore the "noise." Yet, out of today's noise comes tomorrow's new paradigm.[14]

Historically, new ideas and theories generally encountered resistance: some scientists refused even to look through Galileo's telescope. When he taught Copernicus's theory about the planets revolving around the sun, the authorities placed him under house arrest for the final eight years of his life. The discovery of x-rays at the end of the nineteenth century violated entrenched expectations and opened up such a strange new world that eminent scientist Lord Kelvin pronounced x-rays an elaborate hoax.[15] We ascribe Galileo's persecution to the dominance of religious dogma and think that we are more open-minded in modern times. Yet examples abound of data or theories that have taken decades to be heard because they ran counter to the prevailing scientific dogma. Eminent astronomer Arthur Eddington so ridiculed Subrahmanyan Chandrasekhar's theory about the relationship of the mass to the collapse of stars, that twenty years passed before it was accepted.

Geneticist Barbara McClintock worked virtually in isolation for over thirty years, holding to her vision of the complexity of genetic organization that differed from the central dogma of genetics, before her work was recognized. Over the decades, she persisted with minimal funding. In 1983 she received the Nobel prize for her discovery of mobile genetic elements known as "jumping genes."

More recently, an editor of *Nature*, one of the gatekeepers of scientific knowledge, proclaimed British biochemist Rupert Sheldrake's book *A New Science of Life: The Hypothesis of Formative Causation* an "infuriating tract" and declared it "the best candidate for burning there has been for many years."[16] He rejected the book as "pseudoscience" on the grounds that Sheldrake did not delineate the nature or origin of the morphogenetic fields he describes. Although the validity of Sheldrake's hypothesis remains unproved, Nobel laureate Brian Josephson contested the editor's outright rejection. Josephson points out that the phenomena of heat, light, sound, electricity, and magnetism were studied long before there was any understanding of their true nature. In response to the editor's claim that only fully testable hypotheses can be dignified as theories, Josephson reminds him that this criterion bars black holes and general relativity from being considered as legitimate scientific theories. In closing his letter, he advocates a more receptive attitude to new developments as he takes the editor to task:

> The [editor's] fundamental weakness is failure to admit even the possibility that genuine physical facts may exist which lie outside the scope of current scientific descriptions. Indeed, a new kind of understanding of nature is now emerging, with concepts like implicate order and subject-dependent reality (and now maybe formative causation). These developments have not yet penetrated to the leading journals. One can only hope that editors will soon cease to obstruct this avenue of progress, and instead encourage reviews of the field.[17]

Acceptance of data or a new theory depends on whether it fits into the current view of how the world works. Studies on fraud in science show that it is easiest to pass off fraudulent work if it agrees with widely believed theories.[18] In contrast, listening carefully to nature forces science to revise its description of reality.

Denial is a useful psychological mechanism for protecting ourselves from dealing with things that are threatening. We simply pretend they do not exist. But in the long run, denial does not work. By courageously being willing to receive, scientists can see new things—things that have been there all along, like changes in the heavens, that others denied as impossible. Although science by definition explores the unknown, the implications of what we find may be more than people are willing to deal with.

RECEIVING THE MONSTERS OF NATURE

In the desire to impose order, rather than listen to nature, science has ignored the messy stuff, the monsters, the noise of nature. Based on the belief in an orderly universe, scientists valued a theory in proportion to its ability to explain cause and effect. Science strived to make nature more predictable. Over the centuries, the chaotic side of nature was cast into the shadow along with the Feminine.

In the way that scientists may dismiss a data point, or journal editors may suppress a theory, Western science has systematically ignored the chaos that surrounds us in the form of waterfalls, dripping faucets, earthquakes, and smoke rising from a pipe. Instead, science has made the most of reducing nature to sets of linear equations, observing proportional relationships between force and acceleration, matter and energy, electricity and magnetism. Linear equations such as $x = y + 1$ are well behaved: they can be visualized by a line on a graph, taken apart and added back together, and used to predict the outcome of an action or explain the workings of a clock. Linear equations can be solved: for every y there is one right answer for x; for $y = 4$, $x = 5$. This power to explain and predict nature led physicist Pierre-Simon Laplace (1749–1827) to boast that, given the initial conditions and the ability to perform the calculations, he could predict the state of the universe at any future moment.[19]

Meanwhile, a whole class of equations—nonlinear equations—were looked upon as unnatural because they represented mathematical structures that did not fit the patterns of nature described by Euclid and Newton. Mathematicians regarded nonlinear equations as "pathological," forming "a gallery of monsters."[20] These equations were decidedly not well behaved. An innocent-looking nonlinear equation such as $x_{n+1} = 2_{xn}$ cannot be solved to get one right answer because the terms are repeatedly multiplied by themselves (these are known as feedback loops). As a result, a small change in one variable is amplified and has a disproportional, sometimes catastrophic effect on other variables. When you plot points on a graph, nonlinear equations do not form tidy lines or smooth curves—they form loops, recursions, and discontinuities. A line might behave nicely for a while and then suddenly branch into two lines. While the solution to a linear equation enables the scientist to generalize and predict other solutions, nonlinear equations tend to be unruly and individual. Nonlinear systems are unpredictable.

Surprisingly, however, within chaotic systems are layered deep structures of order. This is one of the key discoveries of the science of chaos. While solutions to nonlinear equations are unpredictable and have no single right answer, there is not complete randomness. Although each iteration of a nonlinear equation cannot be precisely predicted, it is constrained within certain limits—like the swing of a pendulum. We can receptively await an outcome or interact with the process, but we cannot predetermine or order a particular result.

When I first started to read about the emerging science of chaos, I was immediately struck with similarities to characteristics that had been ascribed to the Feminine: unpredictability, nonlinear processes, the importance of context, and the inseparable relatedness of the parts to the whole. This brings to mind Gilligan and Belenky and her colleagues' observations that women tend to view knowledge and moral choice within the complex dynamics of a situation—rather than based on a set of rules. In chapter 6 we will explore the advantages offered by nonlinear systems that arise from their fine sensitivity to context.

Curiously, the language of chaos theory borrows words from the world of women and the home. In contrast to the starkly abstract and efficient language of mathematics, concepts in chaos science are described as dust, webs, cups, foam, fudgeflakes, folded-towel diffeomorphisms, smooth noodle maps, curds, and whey. Instead of the ideal world of Euclidean points, lines, squares, and cubes, chaos theory opens up the space between one-dimensional lines, two-dimensional squares, and three-dimensional cubes—using fractal geometry to model the irregular shapes of clouds, mountains, trees, and coastlines. (For example, the coastline of Britain has a fractal dimension of 1.26.) As feminist writers such as Evelyn Fox Keller have shown, metaphors and language shape our thought and affect scientific discourse. Chaos science is shifting how we see the world and, as a voice of the Feminine, is changing science at its roots by changing its language, the language of mathematics.

It is interesting that chaos science emerged from mathematics, which had become our most reductionistic way of looking at the world: biologists reduced life to a set of chemical reactions, chemists explained reactions with chemical equations portraying atomic interaction, and physicists reduced matter to an array of mathematical equations. Once a phenomenon had been described by a set of equations, it was "known." It could be predicted, manipulated, and controlled. Meanwhile, mathematicians increasingly distanced themselves from any applications of their work to the world. Pure thinking types, they deduced everything from first principles using rigorous logical analysis. They believed that mathematics should be self-contained, something all by itself.

Then suddenly, the computer, the most logical of human inventions, pushed the limits of logic and opened a door to the unpredictable. By allowing mathematics to be practiced as an experimental science, computers brought mathematics out of the realm of analytical abstractions, logically proven theorems, and theoretical constructions. Now researchers could play around with nonlinear equations, interact with them, and watch unexpected patterns evolve and change on the screen. They could seek to understand behavior that had been too complex for the usual methods of mathematics. Researchers in climatology, population biology, physics, astronomy, thermodynamics, ecology, economics, physiology, and mathematics came together to give birth to the new science of chaos. However, since it has been popularized, many practitioners prefer to avoid the sensational word "chaos" in favor of the more precise "science of complexity."

The study of fractals evolved independently of the "theory of strange attractors and of chaotic (or stochastic) evolution."

Symbolically, Benoit Mandelbrot's interest in fractals originated when he retrieved a book review from a "pure" mathematician's wastebasket. Out of that discarded book review emerged Mandelbrot's lifetime involvement in fractal geometry. For many long years, his interests were not shared by anyone. Like a squirrel collecting bits of shiny foil, he rejoiced upon finding obscure references in ancient works that fleetingly alluded to his concerns and insights about what was missing from geometry. Invariably, their work had fallen by the wayside and failed to be received. Similarly, Mandelbrot's effort to link the mathematical monsters to shapes in nature met considerable resistance, while abstract shapes were calmly accepted. Pure mathematicians preferred to believe they had transcended the limitations of nature.[21]

Jung observed that whenever an extreme, one-sided tendency (such as the mathematician's extreme emphasis on logic) dominates conscious life, something strange happens. The psyche strives to compensate for the one-sidedness. Over time, the equally powerful opposite builds up in the unconscious, gradually inhibiting conscious performance—until it finally breaks through the conscious control to begin a life outside of the shadow of the unconscious. The opposites can then form a partnership. Jung uses the term *enantiodromia* for the emergence of the unconscious opposite in the course of time. He borrowed the term from Heraclitis to convey the idea that everything changes into its opposite in the course of time, like the cycles of generation and decay, construction and destruction. As examples, Jung refers to the conversion of St. Paul on the road to Damascus, and the transformation of Swedenborg from an erudite scholar into a seer.[22] Now, in a sort of enantiodromia, the ultimately precise science of mathematics has suddenly shifted to focus on the unpredictable, the nonlinear, the chaotic, the amorphous "monsters" of nature. Consequently, the cold, dry symmetries of Euclidean geometry are being enlivened by the startling beauty of fractals.

While it is beyond the scope of this book to explore the full richness of chaos theory, there are interesting parallels between the values and perspectives offered by chaos science and the Feminine—and similar reasons why our Apollonian culture set them aside. Due to the either/or binary logic of the West, if order is good, then chaos is bad because it is the opposite of order. The Romans assigned the confused mass at the beginning of creation to a goddess named Chaos. The Babylonians depicted chaos as a monster, the demoness Tiamat, who is slain by the warrior Marduk in order to establish civilization. Aristotle equated the male heaven with order and immutability, and linked the female earth with change and corruption. In contrast to the Western view of chaos as malevolent and vengeful, the Chinese Taoists characterized chaos (Huntun) as generous.[23] While in the past, complex systems have been perceived as irredeemably chaotic, now chaos science helps us conceptualize them as "rich in information rather than poor in order."[24]

In many ways, I believe chaos science speaks with the voice of the Feminine—not that it directly furthers the feminist cause or has been produced largely by women (to my knowledge, few of the major players have been women), but because it focuses on the relevance of attributes that have been negatively valued in Western culture. Ilya Prigogine and Isabelle Stengers, authors of *Order Out of Chaos: Man's New Dialogue with Nature*, emphasize that "chaos represents not just hitherto unrecognized phenomena but an unjustly neglected set of values."[25] I believe we can gain insight into the value of the Feminine by exploring the constructive roles played by disorder, nonlinearity, and noise in complex systems. Like women, nonlinear systems were put aside, as women were relegated to attic offices, so as not to disturb the order of rational discourse. Like women mathematicians who were called monstrosities,[26] the monstrosities brought together under the umbrella of chaos theory are emerging from the shadow of science and making useful contributions to our understanding of nature.

Although nonlinear dynamics were known in the nineteenth century, they were thought to be irrelevant to science because they could not be solved—just as feeling and warmth have been deemed irrelevant. Background "noise" simply interfered with an experiment, just as feeling can interfere with the efficient process of collecting data. Now chaos science is listening to that noise and finding levels of order within the seemingly chance fluctuations. By definition, achieving "order" means classifying, ranking, organizing, and analyzing—all processes of the thinking function. In contrast, feeling appears chaotic because it arises from a personal internal evaluation rather than from logical analysis. Because the feeling function is subjective, it is called unpredictable—yet upon examination, we find it is internally ordered according to value. The feeling function's emphasis on context and the importance of personal evaluation is echoed in chaos science. Rather than statistically averaging the impact a person can make, chaos science reveals that an individual making a small change in the "initial conditions" of a situation can have an enormous impact on the final result. Just as chaos has been reconceptualized as extremely complex information rather than an absence of order, so can feeling be valued as a richly complex process. Similar to solving a nonlinear equation, the feeling function usually does not give us a single "right" routine answer. It is sensitive to the complexities of the unique situation. As we discussed in chapter 3, both the feeling function and Venus, the ocean-born goddess of turbulence, the goddess of love and beauty, have much to offer science.

Throughout the history of science, women have been identified with "Mother Nature" as cyclic, nonlinear, nonrational, and unpredictable. Studies of nonlinear dynamics now show that these qualities are, in fact, vital to life. For example, physiologists now identify chaos with health because nonlinearity in feedback processes serves to regulate and control. In other words, a linear process, given a slight nudge, tends to remain slightly offtrack. A nonlinear process, given the same nudge, tends to return to its starting point. Arnold Mandell, who studies dynamic systems in the brain, says bluntly, "When you reach an equilibrium in biology you're dead."[27] Biological systems are open systems that maintain a *dynamic* equilibrium of flow and exchange with the environment. The masculine Aristotelian ideal of complete order—of everything forever in its place—leads to death. By infusing robustness and flexibility into the traditional linear approach to research, the Feminine enlivens science.

Some enthusiasts of the new science have put it on a par with evolution, relativity, and quantum mechanics in terms of its impact on the physical sciences—and on our culture. Others prefer to keep scientific concepts within the compartment of science and feel uncomfortable when they are extrapolated to philosophical or cultural issues. Such an attitude is curious coming from these researchers because nonlinear systems teach connection. Mathematically, they prove that all levels of dynamic systems interact and cannot be separated into neat compartments. This realization is shifting the foundations of every scientific discipline. In chapter 10 we will explore in more depth how chaos science shatters our reductionistic view of the world as a clock and gives us new metaphors of flow and wholeness such as the waterfall.

Yet as chaos science matures, the original sense of openness and synthesis is hardening into separate fields that do not receive the insights of the other. While the practitioners of nonlinear dynamics prefer to think of themselves as solving practical and technical problems unrelated to the cultural matrix, Nobel laureate Ilya Prigogine extends the implications of self-organizing systems to the biological and social realms. Rather than focus on the order hidden in chaos, Prigogine has been exploring how order and complexity spontaneously arise out of chaos. At a fundamental level, Prigogine challenges the conventional view of the effect of disorder on the universe. The traditional view of the second law of thermodynamics (which proclaims that entropy always tends to increase in a closed system) is that the energy in the universe is dissipating. This view leads to the pessimistic conclusion that the universe will cool to the point where life will be impossible. Prigogine demonstrates how nonlinear systems go through an alchemical process of turbulence and confusion—and emerge with a more sophisticated level of order. In chapter 9 we will explore how intuition dips into the chaos of the unconscious and imbues science with creativity.

By finally listening to some of the noise of nature, by receptively admitting the monsters of nature into the orderly realm of science, we are fundamentally shifting our perspective of the world. Now let us explore the complex interaction between society, the psyche of the scientist, and the objective world of science.

5

SUBJECTIVITY
Discovering Our Selves
through the Experiment

MECHANICS OF PERCEPTION

All my life I have been drawn to discovering new ways of seeing reality. I've always had the sense that there is more to see if I can only remove a veil of illusion, or look in a new way. When I was a child, the only comic book I ever bought was *The Emperor's New Clothes*, where only a child notices that the emperor is naked. My seventh grade high school science fair project used lenses to change the focus of light—one way of shifting our perception. While driving, I play a game with myself where I try to look with fresh eyes at scenery that I see every day in order to see something new, or to see it in a new way.

Deepak Chopra discusses fascinating studies on the mechanics of perception that reveal how we limit our view of the world by our assumptions. In India they train baby elephants by tying them to a green twig with a flimsy rope for a few weeks. As an adult, a twig-trained elephant tied to a big tree with an iron chain will snap the chain (or the tree) and walk away. But when tied by a flimsy rope to a green twig, it does not even try to escape. It learned that is the extent of its universe. Except for a few pioneers, most animals remain stuck in patterns established early in life.[1] In the laboratory, researchers studying perception raised a group of kittens in an environment with only horizontal stripes. As adults, these cats could not see anything but a horizontal world. They were virtually blind to vertical contours and actually bumped into table legs because they lacked the functional neural connections necessary to see vertical stimuli.[2] Such studies demonstrate that visual experience can modify the brain and limit the ability to perceive.

Humans are not exempt from such consequences of accommodation to the environment. Because they lacked the interpretive framework, the Fuegians visited by Darwin could not "see" the ship at anchor in front of them—although they were excited by the sight of the small boats that ferried the landing party of the *Beagle* ashore.[3] Congenitally blind people who regain their sight after a surgical operation do not know how to interpret what they see. Initially they cannot distinguish a triangle from a square, a spoon from a fork, or a cat from a hen.[4] Only after considerable time, practice, and training can these people make visual sense out of the world. In fact, we all learn how to see. Our perception of the world depends on experience and context. In order to make meaning out of an experience, our consciousness must interpret the electrical impulses transmitted from our senses to our brains. Part of our normal adaptation is agreement with others about what we see and hear—this forms our consensual reality. When someone sees visions or hears voices that other people cannot validate, we judge them insane. Objective reality is something we believe we all share.

Similarly, biology students cannot "see" chromosomes the first time they look at a cell under a microscope. The apprenticeship of a scientist involves learning what is a "real" result versus an artifact. Just as naturalists learn to identify patterns in the mud as animal tracks, so do physicists learn to assign significance to particle tracks. In spite of the adage "seeing is believing," any book on optical illusions reveals how prone our senses are to deceptions arising from our habits of thought.

Such unsettling studies emphasize that our sensory apparatus and our interneuronal connections develop as a result of our initial sensory experiences and how we are taught to interpret them. Thereafter, our nervous system only receives what we have been programmed to see, perceiving only those stimuli that reinforce what we think exists. We automatically screen out all but a fraction of the stimuli surrounding us. Psychologists call this premature cognitive commitment. This cognitive reality is a result of conceptual boundaries that we have structured in our consciousness, after which our nervous system serves to reinforce those boundaries.

Yet human sensory receptors perceive the world in just one particular way. When we look at a flower, we see colors ranging from red to purple. But the receptors in the eyes of a bee cannot see red—and they can see ultraviolet wavelengths. When a bee looks at a flower with its faceted compound eyes it sees a pattern different from the simple geometric shape we see. A bat perceives a flower as an echo of ultrasound. A snake senses it as infrared radiation. A chameleon's eyeballs move on two axes, and I cannot imagine how it integrates what it sees. There is no one right way of looking at the "real" world, but rather a vast number of lenses through which we can perceive reality, all coexisting at the same time, some more useful to our purposes than others. We freeze that field of infinite possibilities into a certain perceptual reality, literally, as a result of our cognition, which is a result of premature cognitive commitment.

Even among humans, we have varying acuteness of the sense organs, plus the larger factor of interpretation of the stimuli. In some ways, instrumentation creates a level playing field—but, in fact, not all instruments are adequately maintained and calibrated. Even the use of instrumentation requires skill, and care is needed in preparing samples for examination.

Humans are also subject to peer pressure in the process of perceiving reality. In one study, students at Harvard University were asked to match a line with one of equal length from a group of three other lines. When alone, the students erred only 1 percent of the time. But in the presence of other students who had been coached to choose a much longer line, the students erred in choosing the proper match 35 percent of the time. Even when the length of the two supposedly equal lines differed by as much as seven inches, some still yielded to the error of the majority.[5] Are scientists superior beings who cannot be fooled so easily?

THE IDEAL OF OBJECTIVITY

A basic assumption of science is that there is an objective universe, separate from and independent of the observer, which can be explored by scientific inquiry. Although objectivity has its origins in moral philosophy and aesthetics, the natural sciences have been touted as its fullest realization. Science strives to be a paragon of rationality, uncolored by subjective, personal, or political bias. Scientific objectivity trains the mind to see the world in a detached, analytical way. For example, biologist Ingrith Deyrup-Olsen constantly struggles to be objective so that she remains open to receiving the marvels of nature:

> I think everything I do has a subjective tone to it, and I have to fight that all the time, because I really want to know what the slug is and how it does what it does. I don't want it to be colored by saying, "I'm not interested in that aspect of what the slug does because that's not what I'm working on." I want to keep an open mind. I'm working now mostly on slug mucus, which is a very difficult, very interesting area, and I have to set up experiments to be blind. I ask my colleagues, "Is that what you think? What are you seeing?" because I am prejudiced. I think I know what's happening, and I really have to fight myself to circumvent my prejudices, and in this field it is particularly difficult. I find I'm in there all the time and I have to extract myself.[6]

The intent of science is to see what is really there—objective reality—rather than what we expect or hope to find. Because of its practice of rigorously testing new theories, we look to science to generate and validate new definitions of reality. Experiments can prove or disprove a theory. Harvard University philosopher of science Israel Scheffler describes science as "a systematic public enterprise, controlled by logic and by empirical fact, whose purpose it is to formulate the truth about the natural world. The truth primarily sought is general, expressed in laws of nature, which tell us what is always and everywhere the case."[7] The objective stance eliminates distorting influences in order to get to the truth. Atmospheric scientist Kristina Katsaros finds objectivity a rewarding aspect of science:

> Sometimes there are heated arguments at meetings about how to interpret data. When you have very few facts, fully interpreting them can give rise to three or four interpretations—within the error bars, the uncertainties in the measurements. You get people adhering to one or the other interpretation for a while, and that's not based on fact

because there are not enough facts. Eventually more facts are gathered and it becomes clear what the answer is, and everybody agrees. In the end you have a new result. That's the wonderful thing about science, that you can only find in science. There is a point when there is no doubt anymore. There is usually a lot of emotional stress before you get rid of some former idea. There may be a few crackpots who fight it, but if the evidence is good, eventually all accept it. I think that's wonderful. One of the best things about science is that there are some objective answers.[8]

In order to make shared public knowledge possible and to build a stable structure of reality, science insists on "objective" research that is disinterested, impartial, detached, and egoless. Scientists adopted the ideal of objectivity in an effort to guard against self-interest, wishful thinking, self-delusion, and personal opinion. The certainty of scientific arguments and experimental proofs guaranteed scientists a certain enviable detachment in the eyes of the world. Confident in the objective evidence, the researcher could remain serene in the face of public apathy or contempt.

In her study of high-energy physicists, anthropologist Sharon Traweek observed that they cultivated social eccentricity and child-like egoism as displays of their commitment to rationality, objectivity, and science. Young physicists often asserted their ignorance of human motives, and of anything "subjective," as if that confirmed their dedication to their work. Development of insight into their own motives and actions was thought to be a diversion of time and attention better spent on science. Even showing a concern with getting along with other people was considered "somewhat unscientist-like."[9]

The motive of "knowledge for the sake of knowledge" is believed to immunize scientists from infection by personal or financial motives for finding a particular result. Historically, objectivity also served to separate scientists from the domination of the church and the pull of politics. A lack of interest in lying lent credibility. In order to ensure objectivity and compensate for the residuum of error in the individual, the practice of science mandates experimental controls, double-blind trials, randomization of experimental subjects, corroboration of results by other researchers, and peer review.

At the institutional level, however, the consensual worldview is imposed on research through the peer review system. This orthodoxy limits and frustrates many scientists, like biochemist Patricia Thomas, whose vision takes them beyond the conventional views:

> To get funding, you have to propose something that fits within the existing body of information. But the existing evidence is circular, and questions like "What is the cause of cancer?" are not being solved. What you need is to take the quirk in the data, the thing that doesn't make sense, and explore it—ask "Where does this come from?" And we will not give people like that money. I submitted two grants, one for something that is totally doable, but basically rather uninteresting and not really directed at the nature of the disease. The second grant took a more searching approach, and said, "I have this interesting observation. I don't know if it'll pan out, but then no one knows what will solve this problem. This different approach can give us a whole new way of looking at the disease." The reviewers said, "This is very interesting, but you can't show that it will come to anything." So while the doable but uninteresting project was funded, the exploratory approach was assigned the lowest priority.[10]

As a gatekeeper of scientific knowledge, peer review provides an organized system for evaluating scientific work, awarding research funds, certifying the correctness of procedures, establishing the plausibility of results, judging the publishability of manuscripts, and conferring honors. By giving a stamp of approval to new knowledge, peer review makes the claims more credible to the nonscientist. But the peer review system is also plagued by cronyism, elitism, and conflicts of interest—and it stifles innovation. In a 1982 survey of grant applicants to the National Cancer Institute, 60 percent of scientists held that reviewers are reluctant to support unorthodox or high-risk research; only 18 percent disagreed with that assertion. Respondents wrote comments such as:

Anything novel had to be bootlegged.

One must never say anything new in a grant application.

The proposals that get funded are generally the most boring and mundane.

Reviewers seem to like a definite plan of attack, which they call "focused."[11]

In this area, the public can impact the direction of research by supporting the exploration of novel approaches, such as the effects of stress or electromagnetic pollution on health. Recently, public pressure caused the Congressional Office of Technology Assessment to look at alternative cancer therapies in the U.S. This turned out to be the most politically controversial project they had ever undertaken.

In spite of the ideal of objectivity, science is also heavily influenced by who pays for it—primarily government and industry.[12] The areas studied, the questions asked, and the decision to publish or withhold information are usually dictated by the funder. Industries have their trade secrets, businesses and universities protect commercially applicable knowledge with patents, and the government assigns top secret classification to sensitive information to protect our national security. But this is nothing new. Throughout history, science has often served the builders of empires—financial and political.

Because of the powerful authority of science, many people use scientific evidence as proof to substantiate their side of an issue. For example, in the case of the Exxon Valdez oil spill in Alaska, research to assess environmental damage has been funded by Exxon, the state of Alaska, and the federal government. Jerry Galt, a physical oceanographer with the National Oceanic and Atmospheric Association (NOAA) involved in the emergency response to oil spills, observed that "a major oil spill is quickly followed by a large lawyer spill and then by a massive money spill [to assess the environmental damage]."[13] In preparing for litigation, each side funds research to support their side of the case, documenting how little or how much damage was done. But until the case settles, the research is kept secret—usually for several years. Neither side wants information published that might weaken their case. In one lawsuit, the settlement stipulated that all the research remain unpublished, thus hindering the reaction to future spills. In trying to gather information so that he can make intelligent choices in responding to emergency situations (rather than inflict more damage in the response), Galt feels frustrated by the "spin doctors" who take the facts and twist them, using scientific information in a partisan way. In cases like this, subjective interests impede good science.

CONSTRUCTED KNOWLEDGE

Does a researcher miss something by being "too detached?" As opposed to the objective researcher separated from the experiment by a veil of disinterestedness, Belenky and her colleagues describe constructivists as passionate knowers, "knowers who enter into a union with that which is to be known."[14] Sociologist Barbara Du Bois describes "passionate scholarship" as "science-making, [which is] rooted in, animated by and expressive of our values."[15] Rather than seeking self-awareness so that they can edit the self out of the science, constructivists use the self as an instrument of understanding. Both thinking and feeling, objective and subjective, become aligned in the pursuit of knowledge. Tiring of the scramble for funding, developmental biologist Sylvia Pollack enrolled in a Masters degree program in psychology to prepare for a career change. Even though she enjoyed the science, she said, "It's the environment for doing it in that has worn me out." Meanwhile, she continued her work in the laboratory. As a result of the classes, she found herself growing and changing. Psychological insights and knowledge about herself interacted with her science and enriched her work. She gave herself permission to have fun once again with her science. In a course dealing with creativity entitled "Approaches to Knowledge," the class discussed the perspective that knowledge is created. Pollack reflected:

> All my life, for at least the first fifty years, I believed that knowledge was revealed and that it just sat out there and what you had to do was keep looking and you'd see it. My point of view has shifted radically on that, since I think there are so many realities and we only look at what we want to see. I think it's as true in science as it is in the arts or anywhere else. It's changed the way I do science. I used to be quite passive about it in the sense that I'd decide what was interesting and do experiments and look at the data. I feel much more interactive with my science now.[16]

Rather than waiting for the full story to reveal itself from the data, she feels freer to make a story out of the existing pieces—a story that "works for her" while coherently fitting all the other bits of knowledge about the system—knowing full well that there are "umpteen" other interpretations that could be made. This approach gives her a sense of personal empowerment as a creator of knowledge, not just a recipient.

As opposed to taking a skeptical, critical, adversarial stance in the battle of ideas, Belenky and her colleagues' constructivists are much more likely to replace doubting with believing as the best way to "get the feel of a new idea."[17] In a similar way, Marilyn Ferguson uses the term "experimental belief" to describe an attitude of openness to listening to new concepts.[18] The art of listening, of being open, of allowing something to enter without immediately intercepting it with our thoughts and interpretations, requires a temporary suspension of disbelief. The more we are aware of our own thoughts, the more authentically we can listen to ourselves— and the more open we can be to nature, and to new perspectives of reality.

Experimental belief provides a tool for challenging premature cognitive commitments. It does not commit the listener to belief— but rather than defending against the new idea or forcing it to fit an old conceptual framework, the listener experiments with belief, asking in a self-reflective way: if this were true, then what? What does it imply? What are the consequences of this idea? Why do I find this concept so unsettling? What would it mean to the way I see the world? How does it affect my science? The experimental believer tries on an idea by taking it inside and seeing where it grates against or integrates with previous beliefs.

In the procedural knower (one who uses systematic analysis based on reason and procedures), a sense of authority arises primarily through identification with the power of a group and its agreed-upon way of knowing, such as the scientific method. Although this promotes the power of reason and objective thought, it often comes at the cost of a person's authentic, individual voice and awareness of self. From a process of intense self-reflection and self-analysis, such as Pollack describes, emerges a new way of thinking. Rather than identifying with being a scientist, constructivist researchers reflect on the contexts that confine and define them, and bring their inner experience to their work. They weave together the strands of rational and emotive thought, integrating knowledge that feels intuitively important to them personally with knowledge acquired from others. Belenky and her colleagues write:

We observed a passion for knowing the self in the subjectivists and an excitement over the power of reason among procedural knowers, but we found that the opening of the mind *and* the heart to embrace the *world* was characteristic only of the women at the position of constructed knowledge.[19]

Constructivists realize that answers to questions will vary, depending on the context in which they are asked and the frame of reference of the researcher. Examining the basic assumptions and the conditions in which a problem is cast becomes as important as collecting data. It does little good to collect reams of data if we are asking the wrong question or if the question is out of context. The constructivist understands that theories are simplified models of reality, not absolute Truth.

In my scientific training I learned how to choose experimental methods and design experimental controls to minimize bias in my results. But no one ever discussed how my unconscious beliefs and assumptions about the world influenced my science. In fact, our beliefs about the nature of reality determine where we look for answers as well as how we interpret the data. For example, neuroscientists believe that the only place that memory can reside is in the brain. This reflects the Western materialistic belief that the only things that exist are those that can be measured with our five senses or with instruments as extensions of these senses. Damage to a part of the brain is interpreted as damage to the storage unit.

On the other hand, a Hindu scientist who believes in reincarnation would approach the problem with an entirely different worldview and would look in a different place to find where memory is stored. Since some people report remembering past lives, the Hindu's theory would have to account for how these memories are retained after the death of the body. This scientist may theorize that memory is stored in an electromagnetic matrix that becomes associated with the body and theorize that the brain is the receiving unit for information stored in that matrix. In this world view, the brain could function like a television receiving a signal. Particular areas of the brain may respond to different types of signals. In this case, damage to specific areas of the receiver would have the same effect as damaging the storage unit itself.

Others may look to the body as a place where memory is stored. Psychotherapists and massage therapists are aware of memories that seem to be stored in the muscles and fascia of the body. For example, when one client's clenched jaw was released, it triggered her recollection that she was sexually abused as a child—something she had "forgotten" for thirty years. Ballet dancers depend on body memory to recall steps from a dance learned years ago. There are phone numbers that my mind has forgotten, but that my fingers can recall on a Touch-Tone phone.

Because of our premature cognitive commitment to the Western materialistic approach, no experiments are even devised to look for memory in an electromagnetic matrix because it is outside our collective worldview. Such a theory would not get grant support because it seems too "off the wall," weird, and implausible. Such work is labeled "pseudoscience" and neuroscientists risk their reputations to pursue such a line of investigation.

EARLY COSTS OF OBJECTIVITY

Up until the mid-nineteenth century, most observational and experimental reports were narrated in the first person singular, and personal traits such as skill in manipulating finicky instruments or seasoned judgment were prized. While sometimes too much weight was given to the social status of the observer, the competence and integrity of the observer figured heavily in judging the worth of the information. Researchers were proud of their hard-won qualifications and alert to minute differences in the qualifications of others. Although scientific academies published proceedings, the real scientific communication took place through extensive, often lifelong, correspondence with peers. Although they might never meet, their letters often lapsed into intimate exchanges, with personal revelation woven through their scientific findings.

During the middle decades of the nineteenth century, the scale of scientific labor and organization grew and complexified. Better postal systems, railways, new instrumentation, new methods of data analysis, standardized units, and uniform definitions all aided the exchange of information. The increase in uniformity facilitated scientific communication across boundaries of nationality, training, and skill. Increasingly impersonal and formal communications replaced scientific memoirs and personal correspondence. But the demands of communications spanning greater distances and more researchers undermined both trust and skill.

Scientists, like manufacturers, sought cheap labor in order to increase their productivity. They divided research projects into simpler tasks to minimize the educational qualifications required of their assistants. This resulted in "the interchangeable and therefore featureless observer, unmarked by nationality, sensory acuity, by training or tradition, by quirky apparatus, by colorful writing styles, or by any other idiosyncrasy that might interfere with the communication, comparison, and accumulation of results."[20] In place of trust grew a strong preference for mechanized observation and methods, with an ever-more exclusive focus on the communicable. The ineffable qualities of skill and judgment were expensive, rare— and could not be communicated. Researchers turned to statistical methods to standardize their results. Consequently, much valuable information disappeared from the research report. As the style became more rigid, writers no longer mentioned the particulars of an experiment: whether the staining solution had been accidentally left in the sun, whether an instrument was on the fritz, what books the observer had read that influenced his or her thinking—such as the fact that both Darwin and Wallace read Malthus's *Essay on the Principles of Population* while they were formulating their concepts about evolution based on natural selection.[21]

The adoption of standard techniques, instruments, and communications helped scientists reach a consensus about reality. Consensus also arose by tacitly agreeing to limit the attention of science to those phenomena that can be brought under good experimental control. Some phenomena were too variable and capricious to withstand the rigors of traveling from one laboratory to another. Scientists solved the problem of replication by turning away from miraculous incidents, remarkable reports, rare events, and mysterious happenings out in the world. Beginning in the mid-seventeenth century, science increasingly focused its attention on well-behaved phenomena inside the lab or logical theoretical proofs on paper.

The objective came to be identified with the mechanical.

Quantitation replaced narration. The interchangeable observer replaced the testimony of a trusted colleague. Independent of the observer, the numbers could speak for themselves and stand on their own. Historian of science Theodore Porter refers to quantitative objectivity as a "technology for dealing with distrust," where quantification provides authority in the form of "power minus discretion (as opposed to power plus legitimacy)."[22] As a strategy for coping with distrust, quantitative objectivity represented the loss of yet another level of connection and relationship. More of the personal, the subjective, slipped into the shadow.

As a substitute for trust, objectivity holds prestige in a pluralistic society. While quantification imposes an order on hazy thinking, it also gives license to leave out much of what is difficult or obscure. Statisticians can average away everything contingent, accidental, inexplicable, or personal. In place of the fine craft skills of the researcher, the quality of an experiment came to be measured by statistical significance levels. For example, a 2 percent statistical significance level (reflecting that the result would occur by chance alone less than two times out of one hundred) in a psychological study would lead readers to accept the conclusions of the paper. In the absence of relationship, public standards and formal knowledge replaced personal judgment and private wisdom. While modern scientists personally know many of the researchers within their narrow field, technicians and graduate students do most of the "hands-on" work today.

As a tool to bridge distrust, objectivity served to create a consensus reality—but at a cost of the variable and the particular. In the quest to make knowledge more reliable, scientists focus on the external world of facts, data, objective observation, and detached experimental procedures. In so doing, Western science defines reality as what exists in the outer material world, and regards the inner world as mere fantasy and imagination. As a defense against superstition, this philosophy of logical positivism permeates modern science. In most fields of psychology, for example, only measurable behavior can be discussed scientifically. Most psychologists dismiss attempts to study the inner, subjective experience based on self-reports. One of the exceptions is the Institute of Noetic Sciences, which is dedicated to studying the mind and human consciousness, with emphasis on the subjective experiences that science has thrown out.[23]

However, creation of consensus reality through objectivity does not necessarily equate with "truth" or "reality." While some phenomena have lingered in the shadow of science, they are no less real for their lack of standardization. Although these phenomena were set aside for valid reasons at the time, those scientists who now attempt to study them are disdained and ridiculed. But slowly some fringe fields, such as nonlinear equations in chaos science and biofeedback in medical science, are slipping back into the mainstream as science finds ways to explain, handle, measure, or quantitate the variable and particular.

THE LIMITS OF OBJECTIVITY

To what degree is objectivity actually possible? The view of the scientist as hermetically sealed within the laboratory has been challenged by many scholars. Sociologists have shown how science is socially constructed. Feminists have revealed the masculine bias of science. Historians of science have demonstrated how philosophical, religious, cultural, political, and economic values can shape scientific judgment. Thomas Kuhn discusses at length the social and intellectual matrix that envelops scientists and their theories, and reveals how paradigms shift during scientific revolutions. But for many scientists and scholars, objectivity—the certainty of hard data—anchors science in reality. For example, philosopher of science Israel Scheffler worries about objectivity coming under attack:

> That the ideal of objectivity has been fundamental to science is beyond question.... The extreme alternative that threatens is the view that theory is not controlled by data, but that data are manufactured by theory; that rival hypotheses cannot be rationally evaluated, there being no neutral court of observational appeal nor any shared stock of meanings; that scientific change is a product not of evidential appraisal and logical judgment, but of intuition, persuasion, and conversion; that reality does not constrain the thought of the scientist but is rather itself a projection of that thought.[24]

If science is not completely objective, it leaves us with the queasy feeling of a slippery, relative, subjective world. Without a handle on what is real, we fear the solid world will slip through our fingers like sand.

The 1989 booklet "On Becoming a Scientist" published by the National Academy of Science's Committee on the Conduct of Science recognizes that scientific knowledge emerges from an intensely human process. It acknowledges that much of the large body of knowledge used by scientists in making decisions "is not the product of scientific investigation, but instead involves value-laden judgments, personal desires, and even a researcher's personality and style."[25] Some of the values inherent in science include: simplicity, elegance, hypotheses that are internally consistent, and the ability to provide accurate predictions. The Committee recognizes that strong attachment to an idea may often be essential when facing the drudge and disappointments of research. In fact, few practicing scientists find that they or their colleagues fit the stereotype of the purely objective, disinterested observer.

In *The Subjective Side of Science*, Ian Mitroff observes that scientific knowledge itself is attained through an adversarial process. Like lawyers zealously representing their side of the case, scientists advocate and defend their theories. As one scientist observed:

> Bias has a role to play in science and serves it well. Part of the business [of science] is to sift the evidence and come to the right conclusion. To do this, you must have people who argue for both sides of the evidence. This is the only way in which we can straighten the situation out. I wouldn't like scientists to be without bias since a lot of the sides of the argument would never be presented. We must be emotionally committed to the things we do energetically. No one is able to do anything with liberal energy if there is no emotion connected with it.[26]

In his interviews of Apollo moon scientists, Mitroff found that most identified a good scientist as one who was highly committed to a point of view. They admitted to overcommitting in order to be heard, selectively looking for data to fit their theories, and provocatively putting their own work forward in the best possible light. Although each researcher tested his idea and backed it up with data, one scientist admitted: "You don't consciously falsify evidence in science but you put less priority on a piece of data that goes against you. No reputable scientist does this consciously; you do it unconsciously."[27] Mitroff concluded that the commitment of individual scientists leads to the full and exhaustive analyses of alternative possibilities necessary for the efficient attainment of scientific knowledge. Rather than remove bias and commitment, Mitroff argues that our goal should be to understand them better so that we can account for their influence.

At a more subtle level, theoretical physicist Brian Martin discusses the presuppositions, value assumptions, and biases in two technical papers that report on the effect of supersonic transport exhaust on stratospheric ozone.[28] In similar studies, articles published in *Science* and *Nature* reach radically different conclusions. Martin points out the many levels at which the authors select and orient their arguments toward a particular conclusion, rather than presenting a balanced presentation of results. He calls this "pushing" the argument and documents how these researchers selectively supported their conclusions by choosing particular technical assumptions and methods, selectively referencing evidence from the literature, ignoring contradictory findings, dropping qualifying statements, emphasizing dramatic results, and referring to alternative arguments in a pejorative way. Even the language the authors chose reveals a bias:

Paper trying to draw attention to an effect that he believes is important	Paper trying to show that this same effect is not significant
ozone shield	ozone layer
burden of NO_x	amounts of NO
threat to stratospheric O_3	interact with, and so attenuate
permitting the harsh radiation ... to permeate the lower atmosphere	radiation reaching the planetary surface

In rebuttals to Martin's study, both authors emphatically denied they had engaged in "pushing." Like these authors, as long as we remain unconscious of our assumptions and biases, the concept of "scientific freedom" remains an illusion. Before we can be free, first we must be able to see the chains that bind us. We must become aware of our premature cognitive commitments in the form of assumptions and the limitations we accept unquestioned.

While complete objectivity may, in fact, not be possible, it remains firmly established as an ideal to strive for. So strong is this ideal that critics such as feminists advocate awareness of subjective influences in order to make science *more* objective. In other words, consciousness of our biases will improve science and make it more reliable.

In discussions of subjectivity versus objectivity, many argue the extremes. While I agree that a totally subjective approach to science is not useful because it does not permit knowledge to be shared, I am interested in exploring the "excluded middle." What if we look at the objective and subjective realms as permeable to each other— where they interact and inform each other? What if the objective view of nature is consciously informed by the personal, and the personal is grounded by the object? What are the benefits of a personal connection to knowledge? From the perspective of the Feminine, all truth is in context. But like chaos, the many different truths are constrained within limits. Although we may never know the Truth, we can explore the diverse aspects of reality, circumambulating the Truth—drawn to it as if it were an attractor, the way a pendulum revolves around a central point, passing through different viewpoints, coming closer to the center with each revolution.

Quantum theory in physics and chaos theory in mathematics present a number of theoretical and practical limits to objectivity. Once we acknowledge the theoretical and practical limits of objectivity, then we can explore how subjectivity can constructively play a part in science.

QUANTUM THEORY AS A VOICE OF THE FEMININE

Just as chaos science transformed mathematics, quantum theory represents an enantiodromia in physics. The advent of quantum theory overturned the hardest of the sciences, the science that epitomized the reductionistic measurement of objective reality. Instead of a solid universe of objects observed by the detached scientist, quantum theory reveals a holistic network of interconnections. Although the physicists involved in unveiling the quantum world

actively grappled with what this means philosophically, most physicists today use the equations without considering the implications
on the nature of reality at the macroscopic level. Nevertheless, the
equations of quantum mechanics touch our daily lives in the form
of nuclear power, transistor radios, microcomputers, digital
watches, lasers, and televisions. Without an understanding of
quantum mechanics, it would not have been possible to interpret
the x-ray diffraction data that led to the discovery of the DNA
double helix.[29]

As in chaos theory, qualities relegated to the Feminine have
emerged in quantum theory. While an in-depth discussion of quantum mechanics exceeds the scope of this book, we can explore how
various facets of the Feminine are represented in the quantum
universe. Some of the hallmarks of quantum theory are Heisenberg's uncertainty principle, the inextricable relationship between
the observer and the observed, and wave/particle duality (complementarity—to be discussed in chapter 6). Heisenberg commented,
"The ontology of materialism rested upon the illusion that the kind
of existence of, the direct 'actuality' of the world around us, can be
extrapolated in the atomic range. This extrapolation is impossible,
however."[30]

The most dramatic demonstration of nonlocality and non-
separability of nature comes from experiments performed by Alain
Aspect's team at the University of Paris-South in 1982.[31] The
Aspect experiment tells us that particles that were once together in
an interaction remain part of a single system and respond together
in further interactions. Aspect's team demonstrated that if you
change the state of one particle in a pair of separated particles that
were previously together, the other particle instantaneously
changes—even though it is further away than communication by
the speed of light would permit it to know. In other words, there is
a connectedness. Unlike electromagnetic signals passing between
the two particles, there is a "knowing" between the two particles
that is communicated or is known to the two particles. Extrapolat-
ing, that means there is communication between all particles, all
parts of the universe, because presumably all parts of the universe
were in connection at the big bang. Since everything that surrounds us can theoretically be traced back to the big bang, all of

humankind and all of nature are part of the same system. So the universe is connection.

According to Heisenberg's uncertainty principle, the actual properties of objects can no longer be separated from the act of measurement and thus from the measurer. While it is possible to measure the position or the momentum (its speed and direction multiplied by its mass) of an electron, we cannot measure both. The act of locating the electron with a gamma ray knocks it out of its path and changes its momentum. But quantum theory goes deeper than that. It is not simply that we cannot devise a small and gentle enough probe that it will not knock the particle off its path. The uncertainty principle states that we cannot know. According to the fundamental equations of quantum mechanics, there is no such thing as an electron that possesses both a precise momentum and a precise position. This implies that not all the properties of a system can be known exactly. If a certain property is measured precisely, then another will become uncertain. Max Born, another of the creators of quantum theory, wrote that "no description of any natural phenomenon in the atomic domain is possible without referring to the observer, not only to his velocity as in relativity, but to all his activities in performing the observation, setting up instruments, and so on."[32] Like chaos theory, quantum theory affirms the fundamental unpredictability of nature (often projected onto the Feminine) so that we can only talk in terms of probabilities.

To be uncertain, like the stereotypical woman constantly changing her mind, has been equated with weakness and has been relegated to the Feminine. But from another perspective, uncertainty gives freedom. Something that is uncertain cannot be controlled. Women raising children, or people living close to nature, live daily with uncertainty. In fact, the most we can hope for is to carefully define our limits of certainty and acknowledge what we are uncertain about. When we consciously acknowledge our uncertainties, rather than passing over them, we are more inclined to be open to other methods of proceeding. Uncertainty opens up options. Rather than rely on planning and highly scheduled efforts to control moving from experiment A to experiment B, acceptance of uncertainty frees us to incorporate feedback and revise our movement and direction as we set off in a general direction. Informed by our uncertainties, we can use feedback to resiliently respond to each situation uniquely, flexibly adjusting our process as we go, negotiating the multiplicity of variables and complexities like explorers in unknown territory—as indeed we are.[33]

While quantum theory implies that reality can never be precisely described, the effects initially seemed limited to the quantum level. Physicists continued to rely on Newtonian mechanics to calculate macroscopic events, such as the motion of a billiard ball. Now chaos theory reveals how small uncertainties and fluctuations at the quantum level can be magnified by iterative processes (like compounding interest in a bank account) until they have a substantial effect on the macroscopic world. A quantum fluctuation can bubble up to the level of everyday reality in a way described by the proverb "For want of a nail the shoe was lost; for want of a shoe the horse was lost; for want of a horse the rider was lost; for want of the rider the war was lost." This type of amplification has been most dramatically observed in weather. (Ever since I moved to Seattle, I have been convinced they send weather forecasters to the Pacific Northwest to learn humility.)

SUBJECTIVITY IN CHAOS THEORY

Science is based on observation and measurement. To the purely materialistic scientist, if one cannot detect something with one of the five senses, or instrumentation that extends the range of those senses, the thing does not exist. Measuring something establishes its reality. Science does not "believe in ghosts" because no one has been able to measure a ghost. Yet we can now measure invisible things such as electricity and magnetism that seemed to be in the realm of magic to the original members of the Royal Society of London. As instrumentation becomes more sophisticated, more and more precise measurements can be made. But does greater precision provide more correct descriptions of nature?

Chaos theory also reveals the interdependence of the observer and the observed. At the macroscopic level, theorists such as Mandelbrot challenge the objectivity of quantitative measure, one of the foundations of science. He writes, "The concept of geographic length is not as inoffensive as it seems. It is not entirely 'objective.' The observer inevitably intervenes in its definition."[34]

Even something as simple as measuring the length of a coastline depends on the perspective of the observer. Unlike research done to prove the superior intelligence of men over women, the measurement of length of a coastline should be a straightforward process, free of any emotional entanglements, political pressure, or unconscious biases of the investigator. Yet, upon closer examination, there are a multitude of decisions investigators must make before embarking on a project. The investigators must decide upon the degree of accuracy necessary and choose the tools of measurement, both of which reflect the intent of the measurement.

What purpose will the information serve? Is it for the construction of a road along the coast that bridges streams and valleys? In describing a hike along the beach, the coastline becomes longer as the hiker negotiates boulders, bays, and inlets. To an ant tediously traversing the pebbles and rocks along the beach, the coastline becomes almost infinitely long. Where does the land begin and the ocean end? Do we make the measurement at high tide or low? The more precise the measurement, as we carefully take into account the crevices in the bits of sand, then the circumference of the molecules making up the sand, the longer the coastline becomes. But does increased precision give a more correct answer? We must answer no, because then all coastlines would be the same infinite length, and that doesn't make "sense," nor is it useful. All measurements must be made within a context. Again, the standpoint of the researcher is intimately linked to the results obtained.

DISCOVERING OUR SELVES THROUGH THE EXPERIMENT

By orienting itself to the object and objective data, science takes an extroverted view of the world and is biased against introversion, where the subject is the prime motivating factor and the object is of secondary importance. By overvaluing our capacity to objectively apprehend the world, we repress the important contributions of subjective factors, labeling them "merely subjective."

Being an introvert myself, my interest in science revolves around what it tells me about new ways of seeing reality. By challenging my assumptions about the world, I discover how I limit myself. Expanding my views of reality allows me to expand my potential. It takes an act of will to "see," to overcome the hypnosis of expectation. I wonder: What if this were true? What does it mean to the way I see the world and live my life? As I learn about the world, I learn more about myself. As I learn more about myself, particularly through Jungian analysis, I learn to take back the projections I cast upon the world.

Projection means "the expulsion of subjective content into an object,"[35] like projecting a picture onto a screen. We use this mechanism to rid ourselves of painful, undesirable, incompatible, shadowy parts of ourselves. And we also project golden parts of ourselves that we have not owned. Projection serves a useful function in the development of consciousness and self-knowledge. By projecting the unexpressed contents of our psyches, it makes them easier to see than if they remain inside. But our projection onto an object does not mean the world is mere fantasy, unrelated to reality. Jung describes the relation between the objective and subjective worlds as follows:

> In the making of scientific theories and concepts many personal and accidental factors are involved. There is also a personal equation that is psychological and not merely psychophysical. We see colours but not wavelengths. This well-known fact must nowhere be taken to heart more seriously than in psychology. The effect of the personal equation begins already in the act of observation. One sees what one can best see oneself. Thus, first and foremost, one sees the mote in one's brother's eye. No doubt the mote is there, but the beam sits in one's own eye— and may considerably hamper the act of seeing.[36]

In order to hold the projection, there must be an adequate hook on which to hang it. Then the projected contents become entangled with the external object, so we must learn to distinguish the object from what we thrust upon it. The occurrence of a projection is signaled by frustrated expectations, by the discordance between what we imagine to be true and the reality. We can detect a projection by being alert to reacting with particularly strong feelings of attraction or aversion, affects that are out of proportion to the situation. Until the projection is withdrawn, the relationship with the object is partly illusory and tends to isolate the subject from the environment. Withdrawing or dissolving projections allows a real relation to the world.

The unconscious part of ourselves that seeks expression uses the mechanism of projection to bring us into interaction with it. When a man falls in love with a woman, he projects the feminine part of himself onto the woman and feels he knows everything about her. Some of those qualities are, in fact, present in the woman—enough to hook the projection. The veil of enchantment lifts when he begins to discern the reality of the individual woman from the whole package of qualities he has projected. When we pay attention, projection serves as the first step toward self-knowledge. On the other hand, the refusal to undertake the task of withdrawing the projections has grave consequences. Insistence on maintaining disowned qualities outside of ourselves results in bigotry, scapegoating, and war.

In his study of alchemy, Jung realized that the alchemists were projecting psychic qualities and their internal processes onto the physical substances and procedures in the laboratory. In symbolic and imaginal language, they described the phenomena as if they were occurring outside of themselves in the substances they were manipulating. At the same time, Jung found statements of spiritual and philosophical truths amongst their descriptions of chemical procedures, evidence that the alchemists acknowledged the psychological level of their work.

Similarly, our work—be it art, business, or science—points to our personal issues. As scientists, we project the contents of our individual psyches onto the objects and phenomena of nature we study. So far, discussions of subjectivity have emphasized how awareness of the distorting influences of personal values and cultural values can improve science. Now let us turn that around and ask about the role that scientific research plays in the personal growth of a scientist.

While I was working on this chapter, I attended a writers' workshop in order to continue to improve my craft. The speaker asked us to take an issue that we care deeply about and weave it together

118

with a personal story. I sat in my chair pondering how my scientific research reflected my personal issues—in the way the alchemists believed that the reactions taking place in the alembic, the alchemical vessel, reflected what was going on in their psyches. According to Jung, the unconscious tends to live itself out in physical acts that bear a symbolic relationship to what the psyche needs to bring into consciousness.

As I thought about my research, my face grew hot. I felt exposed. I realized that my research pointed to psychological issues that have been the substance of my Jungian analysis for the past five years. In graduate school, I studied the effect of anesthetics on beating heart cells in tissue culture—metaphorically, my psyche was pointing to the feelings I had numbed in order to survive an abusive childhood. Later I did research in the area of sexually transmitted disease, which pointed to my issues of sexual abuse as a child. Farfetched? Maybe. But as I sat in my chair making these connections at the workshop, it felt too close to truth for comfort. In protective denial, the scientist in me protested that I was making more of it than is there. But I became intrigued.

I wanted to find out if other scientists found similar patterns in their work. But this presented great difficulty. Like me, most scientists have no idea they are exposing themselves in their science. Many look down on psychologists and disdain them, saying that psychologists are in it "just to work out their issues." The scientists doing hard science—"real" science—believe they are "all together" and doing productive and meaningful work. Most researchers believe they are doing objective science to ferret out the truth about nature, and would deny that they're working on psychological issues in the lab. Their work is about revealing nature rather than being personally revealed. A sense of separateness from the work, devoid of awareness of self or others, gives a feeling of dignity. Most scientists adamantly refuse to admit they are doing anything but important scientific work.

While I do not doubt that they are doing significant science, I wondered if there exists anything such as an untainted experiment. In addition to this veil of denial, most scientists simply have not had the time or inclination for personal reflection and introspection. They are not trained to think in psychological terms and remain unaware of their hang-ups and issues. In our culture, a person in therapy is still often looked upon as "sick" rather than as growing. Such dynamics make a "scientific study" well nigh impossible. However, I did find several scientists who felt comfortable exploring the issue. Since their stories were shared in confidence, touching the most personal and vulnerable parts of their lives, I will not use their names.

One cell biologist began her research career working on spore formation in bacteria, studying what causes bacteria to make the commitment to change—to synthesize a protein coat that provides a defense against adverse conditions. Coincidentally (?), one of her issues centers around when to leave—relationships, jobs, and other situations that have turned sour. She tends to hang onto things too long. We joked about her old wallet whose halves were held together by just a single thread. More recently, she completed a project on the effect of growth factors on cell adhesion. We talked about how both research projects followed the same theme of attachment and when to make the commitment to change. Several weeks after this conversation, I talked to her about her new project. She said she planned to study metastasis in cancer, particularly cell motility and what promotes their movement. When I mentioned how it continued the same theme we had discussed earlier, she laughed and said, "That's true. In other words, I'm letting go and moving forward. That's funny! You know that's true, isn't it? I hadn't thought about that conversation we had!" At the same time, she was moving toward leaving the company where she had worked for over eight years.

A scientist studying breast cancer was undergoing her second bout with this cancer in her own body. When she was offered a position in an epidemiological study on breast cancer, she joked that she wanted to do work on prostate cancer—"something she couldn't get."

An immunologist was drawn to science to compensate for her chaotic childhood. To her, science seemed to be the only thing that had any real validity, the only place she could find real truth. In contrast to her unpredictable reality, she felt she could find order and truth in scientific laws. She worked for over a decade in the area of autoimmunity, studying why people mount an immune response against their own tissues. Over the past three years, she has been consciously confronting deep psychological issues. As we talked, she connected her work on autoimmunity with a profound sense of self-rejection she felt for close to fifty years of her life. She said,

> The big issue for me all my life was I never felt like I had a right to live. Period. That's pretty much self-rejection and that's autoimmunity, if you put it that way. It really does fit, because the rejection was more from the inside than anywhere else, which is why I had such a hard time. It was rejection, personal rejection, the rejection of myself that was so hard on me. Plants and animals and people around me never questioned that they should be alive and I questioned it every day as a child. I never really could get hold of it, it was just this overwhelming feeling. It was never intellectualized to where I could work on it because it was so deep, so fundamental, so early. It has taken a big effort to accept and not reject the self. Now I am past that and have decided I have a right to live and I don't have to be so hard on myself. And as I've gone through this, my need to do the work on autoimmunity has just about ended. I'm not that interested anymore. Now I want to do some healing work.

In the course of doing research, certain projects present themselves but are not "interesting" and so are not pursued, while other projects become the center of a life work, such as this woman's research on autoimmunity. The link with the psychological issue gives the psychic energy for the scientific work—a project interests us because it is important to us at a deep psychological level. The personal, the subjective, impels science because we are projecting our personal issues and trying to solve them in the lab. It activates and informs our work. We become the alchemical vessel in which the interaction takes place and, if we are successful, something new emerges.

From her work with a number of clients who are scientists, educational psychologist and Jungian analyst Anne de Vore gives an example of a geophysicist who had been blocked for fifteen years from finishing his thesis on permafrost in North America, where the topsoil never thaws. The analysis centered around his frozen feelings and the frozen women in his life (his mother and his wife). When de Vore made the connection between his thesis work and the frozen people around him, then he could get on with his life. He did not have to do the thesis anymore. Instead, he took a job in the Antarctic Division of his company, living in the "deep freeze" for a year. Later he divorced and married a warm woman.[37]

Sometimes our issues and hang-ups are more obvious to others than to ourselves. One researcher told me of a colleague who was working on the bad effects of vasectomies. Her coworkers joked about her being a castrating woman. She was, in fact, a castrating wife and the jokes touched too closely, enraging her.

While the above examples do not provide conclusive proof that psychological issues motivate scientific research, I hope they will stimulate others to start wondering how their personal issues may be reflected in their own work. You might ask, if you were to look at your research metaphorically, do you see any relationship between it and psychological issues you are working on, or areas in your life in which you are blocked? I also think it applies at the collective level of what science we fund—for example, the big science projects such as military defense, the Human Genome Project, the space program, and particle accelerators.

But does viewing science as a tool for self-knowledge give us better science? My biased opinion is yes, that more conscious scientists do better science, and that both contribute to the evolution of consciousness. If we are conscious of how our research is symbolic of ourselves, we can try out solutions in the laboratory as well as integrate them into our lives. Our inner and outer lives become consciously linked. Through the course of a project, the progressive psychological and scientific developments inform each other. Solutions to technical problems may even present themselves in our dreams. But once we solve the inner issue, the energy seems to withdraw from the topic. The science becomes boring and we seek a more stimulating project.

As an iterative, interactive process, self-knowledge gives us insights into the science, and the science sheds light on personal issues. As we discover more about nature, we discover more about ourselves. If we are aware of the link between the psyche and the experiment, we can ask ourselves different questions when the work seems to be going nowhere. When we are blocked in an experiment or a research program, or when problems continually arise, we can step back and consider whether it may be something subjective. Perhaps the experiment is pointing out a need to shift our perspective both in the personal realm as well as in the scientific research. We can ask, if I phrase this problem metaphorically, what is going on here, what are the underlying dynamics?

Science has helped to lift the veils of illusion that kept us living in fear of natural forces and has given us new ways of seeing reality. Knowledge of the ways of nature helps us deal more effectively with reality and gives us new powers. While many scientists have studied nature to learn more about God, we can also examine our interaction with nature to learn more about ourselves, contributing to the evolution of consciousness. This self-knowledge can help us handle our newfound powers more responsibly—as we too become creators.

─── 6 ──────────────

MULTIPLICITY
Webs of Interaction

In mythology, the many-colored veil of the Egyptian goddess Isis symbolized the creative spirit clothed in material forms of great diversity, the ever-changing form of nature. In this chapter we will examine the bias of science toward hierarchy, simplicity, linear progress, and either/or thinking and how these tendencies influence our concepts of nature. As we circumambulate discussions of the social organization of science and emerging scientific theories, we will explore the value of multiplicity, diversity, complexity, interdependence, and cyclic processes that arise out of the feminine principle of relatedness.

A PREJUDICE FOR HIERARCHY

Traditionally, Western culture has identified cyclic forces with the Feminine, while linear progress has been ascribed to the Masculine. Similarly, women's organizations tend toward circular structures (such as sewing circles), while men's organizations tend toward hierarchical, ladderlike structures. In fact, hierarchy has become so identified with the male organizational style that the *The Synonym Finder* lists "patriarchs" and "men at the top" as a synonyms for hierarchy.[1] In a study of the conversation styles of women and men, linguist Deborah Tannen documents how men tend to use language to protect their independence and negotiate status, while women use conversation to establish a world of connection in which individuals negotiate complex networks of relationship and try to reach consensus.[2] Hierarchical thinking continually forces us to rank one thing or person above another. In doing so, it reduces the value placed on multiplicity.

As an institution formed by men, Western science reflects this bias for hierarchy—both as an organizational principle in the social structures of science as well as in presumptions about how nature is organized. In the hierarchy of scientists, physicists are the most elite, followed by chemists, biologists, and psychologists, with social scientists at the bottom. Even within physics, theoretical physicists command more prestige than experimentalists who get their hands dirty. Molecular biologists rank higher than physiologists. "Pure" or "basic" research carries more status than "applied" research. As one graduate student said, "You're not valid in science unless you're doing pure research, not dealing with the rest of society."[3] Amateur scientists are not even mentioned in the hierarchy of scientists.

Hierarchical structures are useful in situations requiring rapid response, such as in military maneuvers. They can quickly create order and efficiently produce results. The leader organizes people and assigns tasks. Such a structure remains benign as long as the followers participate or obey out of choice. Like Plato's philosopher king, this type of ruler serves the common good and leads the others toward a mutually beneficial goal. But since there is only room at the top of the pyramid for the elite few, hierarchical structures usually foster competition, power struggles (the role of competition and cooperation in science will be discussed in depth in chapter 8), and rule by fear and intimidation.

In the scientific hierarchy, the higher positions are special and privileged; the lower ones are anonymous, interchangeable, replaceable. I have often heard of technicians referred to as "a pair of hands." In research facilities I have observed, the head of the lab has a private office, while graduate students and postdoctoral "fellows" (commonly called "postdocs") usually share an office or cubicle. Technicians may or may not have desks, which represent a place to think, read, and write. Although technicians may do the majority of the hands-on experimental work, they frequently are not listed as authors on papers, but are merely thanked in a footnote for their "technical assistance." No matter how skilled, physical work is rarely valued as highly as intellectual work. When postdoctoral positions are in short supply, young scientists desperate for a job are advised not to accept jobs as technicians. To do so implies a "lack of initiative to be a good scientist" and undermines chances of getting an academic position.[4]

Authorship of papers reflects the hierarchy more than the actual contributions to the work. While the names of graduate students and postdocs are listed on publications, they may be given a subservient position in spite of the fact that the lab chief made little or no contribution to the research. In the eyes of the scientific community, the credit for the work rests with the lab chief. When the work is worthy of honors or prizes, the honor goes to the lab chief. For example, pulsars were first discovered by Jocelyn Bell as a graduate student in Antony Hewish's lab. The Nobel committee, however, awarded the prize to Hewish, not Bell.[5]

In her book *Beamtimes and Lifetimes: The World of High Energy Physicists*, anthropologist Sharon Traweek describes the hierarchies in the scientific communities of high energy physicists in the U.S. and Japan. She reports that the physicists at the Stanford Linear Accelerator (SLAC) see themselves as an elite whose membership is determined solely by scientific merit. They view their hierarchy as a meritocracy, a natural ranking of human talents necessary for producing good physics.[6] Status in the hierarchy determines access to laboratory resources (such as precious time on the most sophisticated detectors and particle accelerators), which in turn determines who can do the "significant" experiments. Because the leader has to prove himself in order to achieve his superior position, he considers himself entitled to make the decisions. He generates the ideas, which the subordinates execute and amplify under his direction. Decisions about scientific purposes are made from the top down, where the leader informs the group how the decision will be implemented. Subordinates only find out what is happening after the leader decides. Each person in the hierarchy patterns his behavior according to that of his superior, observing and listening to those with higher status. It is not considered appropriate to comment negatively on those in higher positions.

In hierarchical encounters, each person emerges either one-up or one-down. As one physicist said, "Everyone is looking for some way to make themselves look better, to make the other guy look worse."[7] Among the SLAC physicists studied by Traweek, the preferred style was confident, aggressive, haughty, and even abrasive. Physicists who were "too nice" were not likely to be successful. Traweek noted that letters of recommendation often said that "even though the candidate is quiet and mild-mannered, s/he does excellent physics."[8] Absence of bravado was considered a weakness for which the candidate had to compensate. Although each group leader privately admitted that postdocs often pulled him out of the fire, all asked Traweek not to mention this to anyone else.

This hierarchical structure separates people from social interactions with each other. Traweek observed at SLAC that technicians, administrators, and physicists tended not to mingle across job classifications. She rarely saw experimentalists in the offices of theorists. Several theorists told Traweek that an experimentalist would probably feel awkward among the theorists, who have more status. The SLAC Women's Organization was one of the few groups at the laboratory that included people from all occupational status levels, from physicists to file clerks.[9]

Science often imposes a hierarchical structure on nature, describing a world that obeys the laws of nature, with man at the top and viruses at the bottom of the organizational chart. We speak of the plant and animal "kingdoms," of organisms being "higher" or "lower." In the realm of theory, different hypotheses are tried and tested until the one correct theory emerges, thus showing all others to be hogwash, aberrations to be laughed at and quickly forgotten.

In order to rank things into a hierarchy, we must first reduce multifaceted and complicated qualities into a unitary thing that can be measured and compared. In this, objectivity serves hierarchy by converting the complex and subjective into a single number, a finite quantity that can be ranked. For example, some psychologists have narrowed the worth of a person's mind to a single number, the IQ. This casts into the shadow many other qualities of mental functioning that are not captured by the IQ test, such as resourcefulness, flexibility, precision, cleverness, motivation, will, intuition, interest, commitment, persistence, canniness, skill, enthusiasm, creativity, discernment, and so on. Assuming that intelligence is a single, measurable, innate thing also disregards the effects of nutrition, fatigue, stress, and psychological maturity on mental functioning.

This preference for hierarchy carries over into how we tend to perceive nature. Feminists such as Evelyn Fox Keller have analyzed the propensity for science to favor rhetoric and theories that posit hierarchical forms of control. For example, language such as "laws of nature" imply laws imposed from above. Scientists frequently speak of phenomena "obeying" these laws—Newton's laws of motion, the law of gravity, the three laws of thermodynamics, the ideal gas law, Faraday's laws of electrolysis, Dalton's law of partial pressures, Ohm's law for electrical resistance, Fick's law of diffusion, Einstein's photochemical law, and so on. As an alternative to laws to which matter must be forever subservient, Keller suggests the concept of order:

> The concept of order, wider than law and free from its coercive, hierarchical, and centralizing implications, has the potential to expand our conception of science. Order is a category comprising patterns of organization that can be spontaneous, self-generated, *or* externally imposed; it is a larger category than law precisely to the extent that law implies external constraint. Conversely, the kinds of order generated or generable by law comprise only a subset of a larger category of observable or apprehensible regularities, rhythms, and patterns.[10]

Keller discusses how this hierarchical mind-set creates theories characterized by internal unidirectional hierarchies, locating control in a sovereign governing body such as a "pacemaker" or "master molecule." For decades, the "central dogma" of molecular biology described DNA as the executive governor of cellular organization, charged with transferring information in one direction: DNA → RNA → protein. According to this dogma, events occurring outside a cell could not affect the genes. But Barbara McClintock's research showed that the function of a gene varied with its position on the chromosome, and required the admission of environmental, or global, effects. While hierarchy exists in nature—chickens have their pecking order, and ant colonies have their queen, drones, and workers—it is not necessarily the most "natural" form of order.

MOVEMENTS IN SCIENCE TOWARD COMPLEXITY AND MULTIPLICITY

The hierarchical structure emerges out of Western dualistic, either/or thinking, and is based on linear logic—the absolute classification of all things. Light and dark are envisioned as being in conflict because our thinking has come to accept the idea that both cannot exist together. But a human being is neither wholly stupid nor totally intelligent, completely decent nor thoroughly rotten. Light is neither purely particles nor only waves. The clear-cut distinctions between humans and animals are rapidly disappearing as we find that animals are also capable of language and toolmaking. While in Western culture, a person is either a Christian, Muslim, or Jew, a Chinese person may see no conflict between applying Confucian, Buddhist, and Taoist principles and rituals to various aspects of life.

This propensity for either/or, true/false thinking has, in fact, limited the technologies that Western scientists choose to develop. For example, U.S. scientists have largely passed up applying fuzzy logic to technology. Although the term "fuzzy logic" has often been used to denigrate women's thinking processes, now the term has been applied to a branch of mathematics that has emerged over the past twenty-five years. With roots in the inherent ambiguities of quantum mechanics, fuzzy logic simulates the vagueness and uncertainty inherent in human thought processes. It represents events as continuous phenomena rather than all-or-nothing choices. Instead of solving problems through a series of yes-or-no decisions (represented as one and zero in computers), computers using fuzzy logic assign numbers that fall somewhere between zero and one. By basing its decisions on generalizations rather than exact measurements, fuzzy logic encodes value-laden descriptions such as slow, medium, and fast. This allows machines to run more smoothly and efficiently. Usual feedback controls, such as thermostats, trigger air conditioners to turn on or off. But an air conditioner controlled by fuzzy logic slows down gradually as the room cools to the desired temperature, resulting in energy savings of up to 20 percent.

Although fuzzy logic evolved in the United States, only NASA and a couple of U.S. companies have used it for practical applications. In contrast, over fifty Japanese companies have enthusiastically adopted it. Lack of acceptance of fuzzy logic has been ascribed to the prejudice of Western scientists for precision and binary either/ or logic.[11] Rather than producing machines to simply follow the programming of its masters, the Japanese have used fuzzy logic to encode machines with the flexibility to interact with their environment. They are using electronic circuits designed with fuzzy logic to make subjective decisions about sharpness, brightness, and color in televisions, to enhance images in camcorders and cameras, to select the optimum detergent and cycle times for washing machines based on machine-measured clothing weight and dirtiness, and to detect changes in water temperature and adjust flows to prevent scalding in the shower. Fuzzy circuits have also been applied to elevators, anti-lock brakes, and subway cars so that the machines accelerate and brake more evenly. In fuzzy-controlled subways, passengers no longer need to hang onto straps to keep themselves from falling when the car starts and stops.

While the Western bias for precision and simplicity has given rise to powerful science and technologies, the story of fuzzy logic shows these are not always the most useful and practical ways of seeing the world. Rather than forcing every situation to fit simple all-or-nothing logic, fuzzy theory gives value to the excluded middle. The Japanese have used it as a mathematical way of handling the feminine demand for context by providing a way to adapt machines to individual circumstances. Such machines have the flexibility to deal with questions whose responses typically begin with "Well, it depends on the situation."

Multiplicity is inclusive and broadening, giving us different ways of seeing, feeling, thinking, and valuing. At a fundamental level, the wave/particle duality informs us that light is *both* particles *and* waves. Such a "both/and" way of looking at the world does not mean creating a mishmash where "anything goes." In fact, it requires more discernment and discrimination to know when to apply the appropriate model or perspective. In order to differentiate each view by context and situation, more—not less—consciousness is needed. In this, the masculine skills of logic and analysis serve the ability of the Feminine to relate to the complexities of the personal situation. The feminine desire for context and relatedness serves the masculine propensity to classify. Working together, they give us a richer and more accurate view of the world.

Quantum theory reflects a shift away from hierarchical laws imposed on static structures and describes more complex and interactive systems. Rather than simple building blocks of nature, it describes probabilities and relationships. In contrast to the mechanistic laws of the universe, quantum theory speaks of principles and effects: the Heisenberg uncertainty principle, the Pauli exclusion principle, Bohr's principle of complementarity, the photoelectric effect.

The aspect of quantum theory known as the wave/particle duality demolished the classical notion of simple, solid, static objects. The basic building blocks of matter dissolved into wavelike patterns of probabilities of interconnections. Niels Bohr, one of the pioneers of quantum mechanics, expressed the relatedness of the Feminine when he wrote, "Isolated material particles are abstractions, their properties being definable and observable only through their interaction with other systems." In other words, there is no such thing as an isolated particle—particles can only be understood as interconnections, sets of relationships forming a complex web that we call matter. Just as the Feminine represents the principle of relationship, so does quantum theory describe the fundamental nature of matter as interconnection and relationship—not a hierarchy of things. Bound together by webs of attraction and nonlocal connections, each part the system is affected by change in another part of the system.

The wave/particle duality reveals that we must observe objects in many different settings in order to grasp their true potential—thus providing a theoretical foundation for the value of diversity without hierarchy. One way of seeing reality (as a wave or a particle) is no more valid than the other—they are complementary views. It simply depends on the circumstances of the situation and how we choose to look at it and measure it. Wolfgang Pauli said of this complementary relationship, "It rests with the free choice of the experimenter (or observer) to decide . . . which insights he will gain and which he will lose; or to put it in popular language, whether he will measure A and ruin B or ruin A and measure B. It does *not* rest with him, however, to gain only insights and not lose any."[12] This also means that the observer and observed are inextricably connected.

In place of the deterministic and reversible laws of nature, chaos theory (or "complexity," as Ilya Prigogine prefers to call it) also describes a pluralistic view of the world. By demonstrating that precise prediction is impossible in complex systems, it loosens the chains of control, the laws that conceptually bind the universe. The science of complexity allows us to see nature as generative and resourceful, abundant and interconnected. In this new vision, matter is no longer passive, but capable of spontaneous activity and self-organization. Since the Feminine has traditionally been associated with matter and nature, studying complexity gives us a new way to envision and value the Feminine. What looked passive and chaotic through the lens of hierarchy, we can now see as inventive and creative.

But over the past several decades, high-energy physicists, the scientists at the top of the hierarchy, have focused on the extremes of elementary particles and cosmology, believing that the real problems of science remain only at the frontiers of our universe. In his lecture "Is the End in Sight for Theoretical Physics?" Stephen Hawking predicted that by the turn of the century physicists might have "a complete, consistent, and unified theory of the physical interactions which would describe all possible observations."[13] Similarly, Leon Lederman, director of the Fermi National Accelerator Laboratory, reflects the physicist's quest for simplicity and elegance when he said, "We hope to explain the entire universe in a single, simple formula that you can wear on your T-shirt."[14]

In contrast to this pursuit of ultimate simplicity, complexity theory deals with the phenomena of the everyday world, the mundane world that has traditionally been relegated to the Feminine— the world of clouds and waterfalls, flowers and mountains, boiling water and smoke rising from the hearth. Rather than imposing a hierarchical structure on matter, or reducing the universe to a single equation, Prigogine has been exploring how systems organize themselves spontaneously. In the absence of pacemakers or master molecules, Prigogine studies the emergence of order out of chaos in systems as diverse as thermal convection currents (such as those in the atmosphere and in the ocean), chemical autocatalysis, the self-acceleration of an exothermic reaction, the life cycle of an amoeba, and social systems. Instead of looking to a "leader" to create order, each of these systems use feedback loops to enhance movement toward a new level of organization. Rather than look to physics to explain the basic structure of matter and reduce the universe to a simple formula, Prigogine looks to biological systems for the motivation and inspiration to understand complexity.

As an alternative to imposing order from above, Prigogine discusses how properties emerge that were not in the original "program." For example, below a critical threshold of energy input into a system, the random motion of individuals remains independent of each other. While some individuals may respond to a shot of energy into the system, their movement is dampened and the system returns to the random motion of equilibrium. But when there is sufficient energy or stress to cross the threshold, the system begins to perform a bulk movement. It organizes itself to form a new level of complexity. The individuals most sensitive to the influx of energy respond to it and then begin to attract others. Together they begin to behave in a coherent fashion. In their book *Exploring Complexity*, Grégoire Nicolis and Ilya Prigogine write about fluids achieving coherence in convection flows (as happens from applying heat to a pan of water):

> Beyond this threshold, everything happens as if each volume element was watching the behavior of its neighbors and was taking it into account so as to play its own role adequately and to participate in the overall pattern. This suggests the existence of *correlations*, that is, statistically reproducible relations between distant parts of the system.[15]

At a recent workshop on artificial life, over 300 biologists, physicists, and computer scientists gathered to discuss a missing element in the theories of evolution—spontaneous self-organization. Using mathematical and computer simulations, researchers have demonstrated the tendency of complex dynamical systems to fall into an ordered state without any selection pressure whatsoever. Using these models, they argue that evolution is not just due to random mutations followed by natural selection, but rather involves a combination of natural selection and spontaneous order. Interestingly, self-reinforcing patterns (known as "attractors") form only if the system has enough diversity.[16]

In contrast to the language of hierarchy, which speaks of domination and control, the language of chaos theory speaks of organization created by "attractors." This provides a radically different model for organizational structures, a way to visualize self-organization as an alternative to the hierarchical structure. In addition to providing a new lens through which to view nature, they also serve as a model for social organization.

Chaos science also presents a model of a world built on self-similarity and symmetry across scale, where scale is a characteristic size, from large to small. In contrast to Euclidean geometry's awkwardness in describing nature's architecture, fractal geometry describes nature's characteristic structures with just a few bits of information that specify a repeating process of branching. A single algorithm (a set of mathematical procedures) describes a structure formed by a reiterating process of fragmentation, like repeatedly repositioning and reducing an image on a copying machine. For example, a snowflake (as constructed by the triadic Koch curve) can be described by the equation $D = -\log(N)/\log(1/r)$, where the number of parts of the object generated (N) is 4, and the similarity ratio ($1/r$) is $1/3$, giving a fractal dimension (D) of 1.26. Such a structure would require thousands of numbers to describe by conventional means.

Branching structures such as trees or blood vessels are found throughout nature. From aorta to capillaries, blood vessels form a continuum. They branch and branch again until they become so narrow that blood cells are forced to slide through single file. The branching behaves consistently from large scales to small. No hierarchy reigns in these systems. Bigger is not better; each scale is an equally important part of the whole.

The webs of interactions in nature have recently been represented in the Gaia Hypothesis as proposed by James Lovelock and Lynn Margulis. Named after an earth goddess, this hypothesis provides a model for global and cellular dynamics. It describes our planet as a holistic, self-regulating system in which the activities of the biosphere are intertwined with the complex processes of geology, climatology, and atmospheric physics.

BOTH/AND THINKING

In forcing us to choose one or the other, hierarchy narrows multiplicity. Through its emphasis on assigning superior and inferior positions, hierarchy misses the richness of diversity. It reduces the marvelous many-colored veil of Isis to black and white. When multiplicity is valued, diverse perspectives complement and augment each other, each lending a facet of the truth, an aspect of reality, an equally valid experience of the world. Each adds another color to the rainbow of life. As Jung said, "Ultimate truth, if there be such a thing, demands the concert of many voices."[17]

The marvelous ingenuity and complexity of nature teaches us that "truth" has many faces, depending upon the perspective of the observer. Even in science, each new truth is partial, incomplete, as well as culture-bound. In contrast to the direct, linear, masculine approach, the feminine process of circumambulation circles around a problem, looks at it from all sides, and sees all of its relationships. By giving us an appreciation for the complexity of even the simplest atom, the Feminine can replace the arrogance of science with a sense of awe and humility.

Acceptance of multiplicity prompts us to ask, how might both these aspects or perspectives be true? Since theories are simply models, mere metaphors of nature, what aspect of reality does theory A grasp that eludes theory B? If other intelligent people support theory B, rather than dismiss them as stupid for getting suckered into a ridiculous theory, we might wonder what we can learn from it. Acknowledging the value of another perspective does not require commitment to it, but rather an attitude of allowing it life, seeing what happens as it grows.

Both/and thinking requires a complex mind capable of embracing all the possibilities. Psychologically this kind of view can be

135

difficult to maintain in an either/or culture because one does belong wholly to any one camp, surrounded by supporters and validated in one's thinking. But many scientists find their experience forces them to accept complexity. For example, Sylvia Pollack, a cell biologist, has come to regard nature as complex and murky—quite unlike a clock or a machine—yet full of possibilities. She is angered by the simplistic reductionistic approach: "In the real world you *know* that things are interactive. I get very upset at people who oversimplify their data and force it into a linear model, especially if it's something I know about and I know it *ain't* linear. I like nice simple things, but I think that a lot of things aren't simple."[18]

Naval architects are engineering specialists who are responsible for assuring the strength, stability, comfort, and performance of the ships they design. One naval architect expressed delight in the challenge of balancing the multiplicity of factors—all important—that have an impact on the design. A large part of the job is being a diplomat because this can be a difficult concept to get across to colleagues who tend to be too detail-oriented. What intrigues this person most about the work is the complexity of boats:

> Boats are fun! They are complex enough to keep me interested because every one is unique. They're really fascinating sculptural forms. The analysis of how they will perform is very elaborate. And then there's so much complexity in designing the machinery to make it go. It has to be a completely self-contained environment—it's like building a sky-scraper, making it float—upright, if you please—and then making it propel itself across the ocean.
>
> In a ship, everything is a trade-off. You want to make it stronger to survive a storm—that makes it heavier, which, depending on where the strength is, can increase stability or decrease it. Every pound that you put into it not only increases initial building costs, but it takes bigger engines and more fuel to push it around for its entire working life. You'd think that you'd want to make a ship as stable as possible, right? Wrong, because stable means that if a wave pushes it over, it wants to come back up. If it has too much stability and comes back up too quickly, people inside won't be able to function on board. All these things have to be balanced. There's no one perfect answer. It's very much an art form.

While colleagues focus on optimizing the hull, or designing the best possible mast, this person sees how each part relates to everything else and aims toward building the best possible *ship*.

During my training in science, I learned that there was one right answer, one correct interpretation of the data. My teachers lectured with the voice of authority imparting knowledge. They presented the current theories as established facts, as "the way things are"—information to be mastered rather than interpreted. Similarly, in her study of the training of physicists, Traweek reports that there was no debate about alternative interpretations of data. Students learned to solve problems according to the conventional models.[19] One experimentalist at SLAC said that he believed that "a successful postdoc had to be rather immature: a mature person would have too much difficulty accepting the training without question and limiting doubts to a prescribed sphere."[20] (The same scientist felt that the social experience of most women and minorities had taught them to doubt authority, and that this kept them from doing well.) While it may simplify the learning process, such an approach to knowledge blinds us to the multiplicity of interpretations.

Developmental zoologist Marion Namenwirth noted that when women lecture about their research, they point out the limitations of the data and acknowledge potential flaws in the experimental design. On the other hand, scientists identified with the stereotype of masculine authority "project an image of impersonal authority and absolute confidence in the accuracy, objectivity, and importance of their observations. By all appearances, they will brook neither doubt nor vacillation."[21] Namenwirth notes that such an authoritative demeanor is antithetical to the hypothetical, incompletely verified, continually evolving nature of science. This veil of confidence hides the intrinsic limitations of any set of data, where the instrumentation and analytical methods are always approximate, and alternative interpretations abound. Unfortunately, when women diverge from the masculine style of presentation and take pains not to overstate their findings, they appear to devalue themselves.

A zoology graduate student expressed frustration with this authoritative style of teaching science. Having attended a women's college where over half her science professors were female, she found she was more open to learning when her professors spoke in terms of potentials or possibilities, of sets of data that pointed one way or another. This style of presentation stimulated her to think about what she was hearing. She says about one professor:

> She was *so* good. She talked about sets of data as pointing at things, and she didn't talk about this being absolute proof or not being absolute proof. She talked about lots of pieces of data coming together and making something more likely than something else. Occasionally she would say unequivocally this proved this. When we talked about mitosis [a process of cell division], she said they give you the impression that everything in mitosis is really regimented—but you watch it and it's not that regimented. It's kind of randomness going to order.

Rather than blindly accepting a body of knowledge, the students of this professor were asked to use their own powers of observation, evaluate the data for themselves, visualize multiple possibilities, try on new ideas, and thus develop their ability to think and discriminate, analyze and make connections. This is just how multiplicity fosters the evolution of consciousness. It also returns a sense of wonder to the world. While hierarchy challenges the few at the top to develop their ability to plan and evaluate, it reduces followers to the status of children who obey orders. Hierarchy breaks the connection between people by separating them into different levels that cease to interact as equals. Like the physicists who kept secret the fact that their postdocs often pulled them out of the fire, leaders cannot afford to be seen as vulnerable. They must be in control. Relating to subordinates as equals lowers them.

Studies of children exposed to different leadership styles have found that those under authoritarian leadership become more aggressive or apathetic. In contrast, children in democratic situations, who took part in deciding the goals to be achieved, worked in an atmosphere of easy communication and mutual help. Consequently, they produced better work, were more friendly and less dominating with each other, and showed fewer instances of aggressive behavior.[22]

ALTERNATIVES TO HIERARCHICAL STRUCTURES

In the pyramid of hierarchy, someone at the top must be displaced to make room for another person eagerly climbing to the summit. In a circular structure, people meet at eye level and everyone inhabits the same level. The circle can expand to include others without displacing someone. But the circle has only one level, and so it can foster sameness and repetition, and has the disadvantage of hindering advancement of individuals. Uniqueness or excellence may be seen as threatening to the harmony of the group. The spiral, on the other hand, embraces both multiplicity and advancement. Any level can expand to include another person as each individual grows. The head of the lab may be at another level, but in a spiral each level is continuous with all other levels. No one needs to be displaced.

Psychological growth comes from making decisions, wrestling with ethical and moral choices, and living the experiments of our lives. In hierarchies, responsibility for all these things is cast upward. People following orders abdicate individual responsibility, obey orders out of fear of reprisals, and set aside ethical dilemmas in the march to get the job done. In turn, those giving orders become separated from the consequences of the action. Operating from an ivory tower of abstractions, it is easy to drift out of touch with reality. In addition, in the name of efficiency and order, the particular and the variable, the eccentrics and freethinkers are suppressed. When something goes wrong, those at the top can claim ignorance and blame those who actually did the deed. Rather than share power, the elite hold and concentrate power and use it to control. The voices of multiplicity are silenced, cast into the shadow. Strange facts are swept under the carpet. As one Nobel laureate said, "Never let a fact stand in the way of a good theory."

Anthropologist Sharon Traweek found that most Japanese physicists were more democratic and less hierarchical, in contrast to the organizational structure at SLAC. They made decisions by consensus. Everyone in the lab felt well informed and thoughtfully considered the issues at hand. Before decisions were made that concerned the whole laboratory, everyone actively discussed the matter and the group leader consulted fully with each of them. The

139

SLAC physicists, on the other hand, complained that "they only tell us what's happening after they decide."[23] It is worth noting that a hierarchical structure that routinely solicits and considers the input of subordinates is possible. Such a structure is modeled by the crew of the Starship *Enterprise* in the television show "Star Trek: The Next Generation," where each group member contributes a particular talent, strength, or expertise. When any member of the crew is missing, the functioning of the ship is impaired but does not come to a halt due to the resources and flexibility of the remaining members.

Rather than the power-over dynamic found in most hierarchical structures, a cyclic structure fosters power as responsibility: power-to, power-in-the-interests-of, and empowering others. One person described power to me as "being listened to," which implies respect. People who are not heard begin to feel invisible, powerless, and hopeless. Through the process of discussion, the personal and particular emerge from the shadow as each person contributes their piece of the truth. Rather than compromising, creative ways can be found of forging solutions to problems. When the group gets bogged down in either/or thinking, locked in debating whether one solution is better than another, it helps to step back to reevaluate the assumptions and intentions, and then create new options. A range of possibilities can often move the group forward and open up the creative process.

One challenge of such an approach is learning to function as conscious individuals within the group, tolerating the uniqueness of group members, and balancing individual needs of each member with the common good. As all the potentialities of each member are brought to consciousness, new levels of functioning emerge within the group.

Making decisions by consensus can be a conflictual, chaotic, tedious, time-consuming process. But because it takes into consideration the collective multiplicity of thoughts, feelings, insights, and skills of the group, it is more likely to produce decisions that last. Each person grows in consciousness as they listen to the perspectives of others with respect; each has the opportunity to develop empathy and compassion for others; and each comes to appreciate the complexities and difficulties of the decision they reach. In contrast, when decisions are imposed on people in the interests of short-term efficiency, they are often subtly resisted or sabotaged. Sociologist Julius A. Roth observed the consequences of what he calls "hired hand research," where technicians have no stake in the intellectual rewards of research, such as authorship of published papers:

> Even those who start out with the notion that this is an important piece of work which they must do right will succumb to the hired hand mentality when they realize that their suggestions and criticisms are ignored, that their assignment does not allow for any imagination or creativity, that they will receive no credit for the final product, in short, that they have been hired to do somebody else's dirty work. When this realization has sunk in, they will no longer bother to be careful or accurate or precise. They will cut corners to save time and energy. They will fake parts of their reporting.[24]

In the book *Betrayers of the Truth*, William Broad and Nicholas Wade attribute much of the fraud and deceit in science to science's hierarchical structure and its reward system. In this system, scientists achieve more status by publishing more papers and bringing in more grant money. They rack up publications by having more technicians, graduate students, and postdocs working at the lab bench. In the process, however, seasoned researchers are drawn further and further away from the hands-on science, and lab chiefs become less aware of the particulars of the research. In order to maintain their status in the hierarchy, they may succumb to putting their name on the papers of their subordinates with little knowledge of its content. Rather than a temple of truth, the laboratory can become "a research mill, a factory for the mass production of scientific articles."[25]

In contrast to the hierarchical, circular, and spiral structures, Grégoire Nicolis and Ilya Prigogine provide a new model based on self-organization generated through feedback loops. The process in which order arises out of chaos, in which systems organize themselves spontaneously around "attractors," is similar to the alchemical process. In *Psychology and Alchemy*, Jung wrote:

> The way to the goal seems chaotic and interminable at first and only gradually do the signs increase that it is leading anywhere. The way is

not straight but appears to go around in circles. More accurate knowledge has proved it to go in spirals. . . . And as a matter of fact the whole process revolves about a central point or some arrangement round a center . . . drawing it closer to it as the amplifications increase in distinctness and in scope.[26]

Trust is required for the spontaneous organization to happen. When trust is missing, we get hierarchy, which cannot tolerate chaos. The desire for control and efficiency short-circuits the natural process of self-organizing systems. A model based on self-organization tolerates a diversity of opinion and skills. Individuals emerge to *facilitate* different group tasks. Leadership rotates depending on the special expertise required for the particular situation. Those with different but cohesive skills take their turn in assuming power as a responsibility. Instead of using power to control, they take it on as duty because they are capable of it, viewing it as an obligation to serve, as a chance to give back to the group. Within this structure, each person develops their own inner authority and a respect for others. Since power is shared, individuals are challenged to work within the group, balancing their own needs with those of the group.

INCLUSIVE LABORATORIES

Last summer I hiked on the pumice plains around the steaming crater of Mount St. Helens with Becca Dickstein. I had not seen her for about eight years. Fifteen years ago we had both started working for the same company on the same day—her with a B.S., me with my Ph.D. I was her boss, responsible for managing a group of five people. Much to my surprise, during her visit she told me that I had been a mentor to her. She said that I was the first person in science to take her thinking seriously, to "think with" her rather than tell her what to do. (Her name appears on a patent for a product we worked on together, just as other technicians coauthored papers with me.) Contrary to the rules of hierarchy, we formed a lasting friendship. She is now an assistant professor at Drexel University, "thinking with" her students and technicians.

Deborah Tannen, in her linguistic study of conversational styles, found that men gain status in hierarchies by telling others what to do and resisting being told what to do. Compliance indicates submission to the authority of the leader. Women, on the other hand, make it easy for others to express their preferences without provoking a confrontation by formulating requests as proposals rather than orders. Framing proposals with "let's" and "we" implies that the group is a community where everyone participates in formulating goals and making decisions. In addition, men typically did not give reasons for their demands, while women gave reasons why the request served the general good.[27] Tannen observed that women feel uncomfortable giving orders and are more likely to ask questions. Looking back on my experience as a scientific manager, I realize I fit that pattern. Although I noticed that my boss and other scientific managers just told their technicians what to do, I wanted to draw out their ideas and interpretations, and begin to empower them to make decisions about how to proceed.

Albert Einstein is an example of a scientist with this more democratic style of management. His assistant Valentine Bargmann observed that Einstein treated those around him as equals. Bargmann said, "It is very important to stress that he worked not on the basis of master and disciple, but whatever we had to say would be taken very seriously and discussed just as thoroughly as if it were Einstein's idea. What he liked was resonance."[28]

Unfortunately, hierarchy has become so entangled with the reward system in science that it is difficult to imagine another system being successful. Some scientists, however, have made small beginnings within their own area of influence. Through such individual efforts, perhaps a critical threshold will be reached and the entire system will reorganize itself. For example, as a supervisory fisheries biologist at a federal laboratory, Cynthia Haggerty acted as a buffer between the people in her lab and the hierarchy above them. She enjoyed working closely *with* the five people in her lab to plan and carry out the research. As the lab became more successful, she spent more time supervising, going to meetings, lecturing, writing grants and quarterly reports, and organizing international symposia. But she preferred doing hands-on research. She observed that this was not typical of supervisory researchers, saying, "Their sign of success is to organize everybody else doing work—and that's one of the reasons that some of that work is not of the highest quality, because the investigator is not in the middle of it."

Haggerty's lab accommodated people with a diversity of life styles, work patterns, and schedules. She consciously selected people to work as part of the team. Everyone had a personal investment in the work, a personal reward for the work beyond their salary: their names appeared on the publications, they presented papers at scientific meetings, and they participated in the process of making decisions about what went on in the lab. One afternoon a week, Haggerty and her group met for brainstorming sessions. Along with everyone else, Haggerty took her turn at maintaining the lab equipment, or going back into the lab at 1:00 A.M. to tend an experiment. Although there were times when boundaries were needed, for the most part they operated as "all just one bunch." In this free, open atmosphere they published an average of five papers per year. People loved to work in her lab. She said,

> My supervisor would come down and notice that we were all happy and laughing, or we wouldn't be there—we'd be out for long lunch hours, or someone would show up at 10:00 to work. But he didn't realize they'd been in the lab till midnight the night before finishing something, so for a 7:00 to 4:00 lab I ran into problems trying to help my supervisor understand our style of work. People enjoyed themselves! I also listened to people, rather than just telling them what to do. When it was time to go to a meeting, if a person had done a whole lot of work on a project, I would send them. I often went because I did need to go, but if I possibly could, I would send them. And when money got tighter I was not always the one who went.[29]

Haggerty actively searched for people with whom to brainstorm and exchange ideas. But she ran into trouble for talking to people outside the chain of command. Her supervisor lectured her over and over about understanding the hierarchy and staying within it. Her supervisor became particularly upset when she talked to the director about ideas for funding. Haggerty said,

> You're supposed to go to your immediate supervisor and if he thinks it's a good idea, it goes to the next person, and the next person. By the time it gets to the top it's diluted and dull. I would never settle down to work with that system. It made no sense to me. What bugged them was

that our lab was so productive and worked so well. They just simply could not fight that.[30]

When there were federal budget cuts that required personnel changes, Haggerty had difficulty letting some people go because each person made a unique contribution to the group. She also felt a sense of responsibility to them.

It was really sad, when you had a lot of momentum and productivity and could see what you could do as a group—and then have to chop off major programs because funding changed. It's very disheartening, really a waste of taxpayers' money when the system bounces around so much that there is not stability for longer-term projects. It was *very* hard for me to see people relocated or let go. I had a commitment to them, and a deep sense of group and connection—they weren't just there pushing test tubes around; they were people and their careers and lives were important.[31]

Environments that welcome multiplicity and develop a web of interaction rather than a hierarchy, foster the growth of the professor as well as the graduate students. By listening to students and respecting their contributions—rather than posturing and acting as though they know everything—professors also learn and grow. Instead of a one-way flow of knowledge and expertise from master to disciple, an interdependence grows. When professors become caught up in the duties of grant writing and administration, it is often difficult to keep pace with the latest developments in fields other than their own. Students with divergent interests and recent training in other disciplines can act to cross-fertilize the research. Less encumbered by dogmas and traditional approaches, they can offer a fresh view. As atmospheric scientist Kristina Katsaros said:

You learn from the students too. They know the newest things—they take other courses and know more about computers than I do, and sometimes take a different approach. They all have different personalities, so we learn from each other that we do things in different ways. By mixing different people you get more aspects of the total. It's very useful to not have everybody in the same mold. Another great aspect about research is that you learn all along. You never stop learning.[32]

Cancer researcher Sigrid Myrdal often got good ideas by describing an experimental question to her husband or children. Because of their unencumbered evaluation of it, they asked questions that she had already set aside. While she had set them aside for a reason, the reasons themselves had since become invalid, but she had not thought to raise the questions again.

Sharing credit for the myriad contributions to research—from students, family, colleagues, and support staff—engenders enthusiasm, spreading a sense of ownership, responsibility, and empowerment. It stimulates a flow of generosity and fosters an esprit de corps—as contrasted to the protective, paranoid, depersonalized environment created by hierarchies that hold tight to power and credit. In the business side of science in hierarchies, tremendous resources are wasted as people protect their positions by assigning blame, scapegoating, and covering their asses—instead of working together to solve the real problems.

EXPANDING THE DIVERSITY OF PARTICIPANTS IN SCIENCE

Up until the present century, science was accessible to any educated person. In many cases it was a hobby pursued out of personal interest. In contrast to the modern hierarchy of scientists as the experts, with nonprofessionals excluded from participating, amateurs made significant contributions to science. For example, the eighteenth-century musician William Herschel ground his own lenses and constructed his own telescopes. Surveying the night sky, he came upon an object he realized was not an ordinary star. It proved to be Uranus, the first planet to be discovered since prehistoric times. His sister Caroline Herschel executed many of the calculations connected with his studies and, on her own, detected three nebulae and eight comets with their telescope.

During the seventeenth and eighteenth centuries, the Parisian salons competed with the scientific academies for the attention of educated people. These informal gatherings in the sitting rooms of private homes combined conviviality with scientific investigation. Run exclusively by women, the French salons served as a major channel of communication, where the noble and nonnoble gathered to exchange ideas. Discussion of science was fashionable, and the male membership in the salons significantly overlapped the membership of the scientific academies (from which women were excluded).[33]

More recently, Louis Leakey selected a former waitress and secretary to head the study on chimpanzees, an occupational therapist to study mountain gorillas, and an anthropology graduate student to study orangutans. Some people questioned his mental health. But Leakey preferred someone with "a mind uncluttered and unbiased by theory who would make the study for no other reason than a real desire for knowledge; and in addition, someone with a sympathetic understanding of animals."[34] For decades, public interest and *National Geographic* supported Jane Goodall, Dian Fossey, and Biruté Galdikas in their studies of the great apes.

Today, the expensive equipment required to do experiments in fields such as molecular biology or high-energy particle physics takes the practice of science out of the reach of amateurs and hobbyists. In addition to its high cost, science has become so complex and specialized that full-time effort seems to be required in order to participate. Most lay people are intimidated by the theory and background required to make meaningful contributions. Without years of education and the proper credentials, they feel they lack the credibility to be taken seriously by professional scientists.

Amateurs are often disdained as unskilled and incompetent. But some people would prefer hiring an amateur to do a job because they do the work out of love and do not cut corners. Professionals, on the other hand, often think they know it all, and lose humility. Too secure in their knowledge, and inflated with hubris, they get caught short.

Unfortunately, in the wake of high-status big science, low-tech research has been neglected. In biology, sequencing the human genome using sophisticated automated equipment draws funding and researchers away from the older fields of organismic biology. Such observational sciences study diversity and multiplicity in nature and do not bring much money into the universities by way of patents or overhead on big grants. In the press to answer the most "fundamental" questions of science, whole areas are easily passed by. A rich variety of nature still remains to be discovered.

Blinded by the glamour of high-tech science, the teaching of mundane subjects such as taxonomy, biological diversity, biogeography, natural history, conservation biology, ethology, and comparative physiology has declined. At a time when the diversity of biological organisms in the world is rapidly being depleted, fewer biologists are being trained to identify them. Scientists such as David Ehrenfeld are concerned about the "de-skilling" of traditional biologists—a draining of skill and practical knowledge about biological diversity and a loss of skilled practitioners to apply that knowledge.[35] In order to assess environmental damage from oil spills, nuclear power plants, the ozone hole, or global warming, the existing web of life must first be understood. This requires more in the way of time and observational skills than it does glitzy equipment. This is one area where amateurs can play a significant role.

Although spectacular breakthroughs in science often require an equally spectacular budget, there are many areas to which amateurs can contribute. For example, amateur naturalists such as "birders" can help track migratory patterns. While professional scientists concentrate on problems requiring sophisticated technology, there is an increasing scope for hobbyists in filling out the wealth of diversity outlined by the scaffolding of theory.

In the field of astronomy, the professionals have concentrated their limited funds in a few expensive facilities. Such large instruments are in great demand. Researchers must apply to allocation panels for telescope time, then schedule months in advance for a precious few days of observing time—and pray for good weather. Consequently, they can address only a few of the most pressing or "fashionable" questions, targeting observations that will quickly yield dramatic results (and publications), predominantly on the fundamental questions of cosmology.

Dedicated amateurs, on the other hand, can assume a role of major importance. For example, while many assume that scientists know how most stars work, we still have much to learn about how stars evolve. In order to understand this process, a rich variety of variable stars must be observed as they change over the nights, weeks, and years. Variable star research is particularly suited to the amateur. Free of the pressure to publish, they can conduct long-

148

term programs using their own modest telescopes. For example, a friend of mine built a rotating observatory in her backyard from schematics in a *Scientific American* article. In her home, her bed forms an oasis in the midst of a roomful of computer equipment and printouts. Collectively, a multitude of amateurs such as my friend can monitor the skies more thoroughly than professionals. They can regularly observe objects such as variable stars over long periods of time.

In their book *Getting the Measure of the Stars*, W. A. Cooper and E. N. Walker[36] provide an introduction to variable stars, explain how to make measurements on stars with a home telescope, and outline carefully conceived programs for regularly observing suitable objects. These authors encourage amateurs to contribute to professional research, and exhort professionals to collaborate with their unpaid colleagues by helping to direct their efforts. This book tells the observer what to look for, and gives pictures of what stars are thought to be like inside—models that are usually hidden away in professional journals. By making such information more accessible and outlining the unsolved problems in the area, this book brings within the realm of amateur observers the ability to make significant contributions to science. In addition, astronomers have formed the Pro-Am Coordinating Committee to explore ways of bringing professionals and keen amateurs into contact. This provides a wonderful model for expanding science such that a diversity of people can participate.

In another area, the nonpolitical organization Earthwatch recruits lay people to assist on scientific expeditions worldwide.[37] Founded in 1971, Earthwatch arranges environmental and conservation studies, archaeological digs, cultural programs, and projects in most major scientific fields. For the cost of travel to research sites, members can "fit a new piece" to the puzzle of nature. This nonprofit organization seeks to preserve endangered ecosystems, explore the world's cultural diversity, and promote health and international cooperation. Since 1984, Earthwatch has provided funding for Biruté Galdikas on Borneo. Volunteers observe wild orangutans in the forest, nurse orphaned ex-captives, catalogue botanical plots, and work in the herbarium.

Another way lay people can participate in science is to fund research through member organizations such as the World Wildlife Fund, which supports research geared to protect threatened species and their habitats.[38] Founded in 1961, the World Wildlife Fund is the largest international conservation organization, and directs over five hundred scientific projects. Membership contributions support projects such as gorilla conservation in Rwanda, studies of the effects of deforestation on migratory birds, and examination of the contributions of wild plants and animals to industrialized society.

These examples show how collaboration between paid and unpaid scientists can be mutually beneficial. With some training and direction, anyone can experience the joy of exploring the multiplicity in nature. Curious volunteers can gather the facts that have fallen through the holes in the scaffolding of broad theory, facts that can force science to shift to a new paradigm. Today, the prevalence of personal computers empowers people with the capability of sophisticated calculations and data analysis that were unavailable even to professionals twenty years ago. Without the constraints of convention and peer review, amateurs also have the freedom to explore the fringes of science as their curiosity and inspiration move them. For example, it was homemaker Lois Gibbs, not a professional scientist, who brought the toxic waste problems of Love Canal to public attention. As with complex dynamical systems, lay people can organize themselves to explore our universe. As people seek to devote their leisure time to increasingly meaningful activities, many can now choose the adventure of science.

APPROPRIATE TECHNOLOGY

Western experts have often been criticized for imposing their sophisticated technology on "undeveloped" cultures. For example, while serving in the Peace Corps in Africa, marine biologist Rebecca Hoff observed how, instead of digging unexciting but functional wells, many projects put in big fancy water systems that quickly became white elephants. No one had asked the women in the village (who were responsible for providing water to their families) what they needed or how it fit into their lives and culture. After a few months, the villagers could no longer turn the spigot and expect water to flow. They needed to buy and import diesel gasoline to operate the pump. Even with fuel, no one was trained to maintain the water system, no one could fix it when it broke, and they had no access to spare parts. So the women returned to carrying parasite infested water for miles on their head. These villages need technology they can sustain without input from outside. Wells were not glamorous, but they were more practical. In another example, Hoff describes the fishing boat built by a German aid project in her village:

> In the fishing village where I lived, they traditionally made boats from dug-out large trees. Others made boats out of planks cut from local timber. Slapped together with galvanized nails, with no overlapping joints, they fell apart in a couple years. Then there was a German aid project that built this incredible boat with screws and fancy construction—but nobody could get that stuff there and no local carpenter was skillful enough to build more of them. So they had this nice demo boat that they could never reproduce or the cruddy one that could do the job that they could make themselves. When the Germans left, the boat sat there. Another monument. The tendency is to intervene in a big flashy way, because that's what we know—and the government of the country wants technology as status.[39]

Women also do much of the agriculture in Africa, but the male Peace Corp workers cannot talk to women due to cultural taboos. As a result, Western aid groups provide machines to help the men, but do little to help women.

These examples reflect the emphasis on fixing a problem without taking into account the context and complexities of the situation. In adapting technology to people's lives, the social and cultural contexts are as essential to the success of the project as brilliant technological ideas. In order for the fruits of science to reach people, experts must learn how to effectively communicate and interact with people, listen to their needs, and work within their system. If someone in the village is not convinced that the new technology will work for them, or if they do not have a substantial reason to change, the project will fail.

All too often, organizations such as the Peace Corps tend to hire Ph.D. experts or engineers when someone with less status but more cultural sensitivity may be more effective. Rather than listening to villagers' experience acquired over the centuries and working *with* these people to help them develop appropriate technology, the aid development groups have the attitude "we're going to teach you." Even with the best intentions to improve the villagers' lives, information tends to travel one way, from the scientific experts to the naive villagers. Forester Katy Gray attended an agroforestry conference that focused on combining agriculture, animals, and trees in a small plot so that farmers with only a little land could maximize their resources. Although the experts said they wanted to teach methods that were more connected with people, they maintained their hierarchical status. Gray expressed her frustration with this prevailing attitude:

> What was so aggravating was it was a big international thing, run by academics and planners and aid development groups. It was just so arrogant. A lot of that stuff was already indigenous, that a lot of these cultures all over the world already had—and nothing was acknowledged that way. It was all, "we have been studying this and we're going to give you these ideas." There were a lot of field people at the conference who didn't speak at all—a real dichotomy between academic planners who did all the talking using big words you had to mentally translate, and then all the field people who were out there getting their hands dirty who were not giving input because there was not really a place for them at the conference. It was so maddening, the attitude that "we'd discovered this really integrated system that can help the Third World."[40]

Rather than approach the conference as an opportunity to learn respectfully from each other, or acknowledge that they were validating the farmers' experience, the experts co-opted the old methods by cloaking them in scientific jargon. This has the effect of devaluing the wisdom gained from centuries of experience through "unscientific" methods of observation and learning by trial and error.

One sees a more collaborative approach in the work of the Green Iguana Foundation set up by biologist Dagmar Verner to explore the feasibility of raising iguanas in the forests of Costa Rica. The population of these reptiles had been declining in Central America due to deforestation and overhunting (iguana meat is considered a delicacy). Due to the success of the foundation's test farm, local farmers now find that raising iguanas is more profitable than cutting down the forests to raise cattle. More iguana meat can be produced per acre than beef, since the reptiles range through a three-dimensional forest habitat. While this approach appears to exploit local wildlife, small iguana farms can help preserve both a local species and the tropical forests. Incorporating them into the economy may be the only way to preserve them as long as a culture finds little value in wild things in and of themselves.[41]

As we have seen, the "truth" has many faces, depending upon the perspective of the observer. Even in science, each new truth is partial and incomplete as well as culture-bound. In contrast to the direct, linear, masculine approach, the feminine process of circumambulation looks at a problem from all sides and at many levels, circling around it and seeing all of the relationships. By giving us an appreciation for the complexity of even the simplest atom, the Feminine can replace the arrogance of science with a sense of awe and humility.

Seeing the workings of the world as cyclic and interactive, rather than simply linear and hierarchical, prompts us to develop a different value system—to value the process rather than seek only the end result. In this sense, *how* science is done is just as important as what science accomplishes. Values of love, attention, and caring contribute to the quality of the process, and in turn influence the resulting product. Now let us explore how nurturing can contribute to science.

NURTURING
A Long-term Approach

AT HOME IN THE LAB

The lab where I spent five years of my life during graduate school was in a new biochemistry building, a monolith of tan brick on the Penn State campus. Built to the latest standards of efficiency, it felt stark and cold to me. Although I understood the practicality of bare cement floors covered only by clear sealant to prevent corrosion from strong chemicals, the cement was unforgiving to my feet. Exposed glass plumbing emptied the experiments from the labs on the floors above us. We speculated on our colleagues' activities as bright yellow liquid sloshed down the pipes over our lab benches. Fluorescent lights hung from the bare cement ceiling, casting a cool light on tan metal lab benches covered with reagent bottles, beakers, flasks, test tubes, and chrome and steel instrumentation. A single window in the corner of the lab provided an occasional glimpse of the outside world as we went about our experiments. My advisor's office was faceless and impersonal. Books and piles of journal articles lined his desk and shelves. His walls were blank, reflecting nothing of his individual taste or life.

In these surroundings I learned that one's personal life had no place in this microcosm of science. Over the door to the building I sensed an imaginary sign proclaiming, "Only things of science may enter here." Unspoken rules said it was unprofessional to hang posters or pictures on the white cinder block walls.

In the experience of a friend of mine, the rules were more explicit. One spring day she put a rose in a graduated cylinder on her lab bench. When her advisor saw this he was outraged and castigated her for inappropriate use of glassware. Flowers in graduated cylinders were simply not done. This was not the purpose of lab glassware. My friend was stunned by his reaction. In banishing the rose, a universal symbol of beauty, love, and the Eternal Feminine, he was also rejecting the Feminine in science. Ever since, my friend has taken care to keep the things of science separate from personal things.

In private protest, I covered the metal cubicle walls around my desk with prints of Sulamith Wülfing's ethereal art. On my bulletin board I hung photos and cartoons, such as Gahan Wilson's image of two befuddled men in white lab coats opening the restricted access door of Lab H to find themselves standing on the edge of the universe, surrounded by stars and planets. One man comments to the other, "It looks as if Bodecker's project has gotten completely out of hand." It seemed to me a good reminder. On the corner of my bookshelf I surreptitiously nurtured a philodendron named Weed. Gradually over my years in the lab, Weed crept beyond the boundary of my cubicle wall and curled itself around bottles of phosphate buffers. I found that comfortable surroundings put me at ease, and art and photos of friends and nature connected me with other important parts of life.

Now, conscious of the masculine influence on the laboratory environment, I wondered whether other women shared my need to personalize their surroundings, to bring warmth and beauty into the lab, to create a nurturing environment. As the home of science, the laboratory reflects many of the values of science; perhaps the science is, in turn, affected by these surroundings. In this sense, the physical structure of the laboratory can be seen as a microcosm of the struggle women face fitting into science, or, on occasion, the expansion of science to welcome the Feminine. Why discourage comfortable surroundings that put a person at ease and thereby reduce stress? Numerous studies have shown people are more creative and productive over longer periods of time when they are relaxed.

Although "decorating the lab" sounds like a trivial issue, the traditionally nonpersonal lab environment reflects the compartmentalization of science itself, where the lab is exclusively a place for collecting scientific data. Anything else is irrelevant, distracting, and inappropriate. In her study of high-energy physicists, anthropologist Sharon Traweek observed:

> The physicists eschew any personal decoration or rearrangement of furniture that would differentiate their workspaces. This great visual uniformity, coupled with the clean, functional grey metal and glass decor of the building, creates a strong impression of stoic denial of individualism and great preoccupation with the urgent task at hand.[1]

Anything personal is considered unprofessional. "Hanging things on the wall is frivolous, a waste of time," says the masculine voice in my head. "Art, family, community, hobbies, or political issues have no place in a laboratory. They get in the way of science. Science is serious business." Opening the door to the personal also may make us feel vulnerable, since our emotional life is more difficult to control than a logical discussion of data and theories. Intrigued to see how others had structured their working environment, I visited a number of scientists in their labs at the University of Washington in Seattle.

Since I had arrived at the zoology building early, I wandered around the halls while I waited for the time of my appointment with Aimee Bakken, a cell biologist. In the halls of this department staffed partly by women, and containing about 50 percent women graduate students, I noticed that the bulletin boards reflected an integration of science with the community: amidst seminar notices were a Salvation Army sign-up sheet to adopt a family, petitions for pro-choice and death-with-dignity initiatives, and a volunteer sheet for a graduate student mentoring program.

Bakken came out into the hall to welcome me, offering me my choice of tea brewed in a clay pot or freshly ground coffee. While she found me a mug, I glanced about her office. All around me I glimpsed signs of Bakken's full and rich life: a picture of her daughter, postcards from friends, a schedule for a music festival next to a schedule for a cell biology meeting. Amidst technical books, journals, and notebooks of data were reminders of the beauty of nature—sea shells, sand dollars, and autumn leaves. A frog-faced clock humorously reflected her work on frog eggs. The side of her file cabinet framed a poster of young Einstein quoted as saying, "Imagination is more important than knowledge."

After talking about her work, interspersed with reflections of how her personal psychological growth has augmented her relationships and her science, Bakken led me into her laboratory. I immediately noticed that the lab was arranged communally by function, with stations for microscopes, electrophoresis, isotope labeling, and so on. The lab felt lived-in and productive. Speckling the walls between shelves of books and supplies were wildlife posters and postcards. A red plastic spider playfully stalked a recipe box of reagents by the pH meter. The plants on the windowsills blended with the rhododendrons outside the lab.

A lab such as Bakken's sets the stage for a different style of interactions between people, as well as reflecting that style. Bakken develops personal relationships with the people who work in her lab that include the whole person, not just the brain and pair of hands performing the experiments. In her soft-spoken voice she observed:

> I seem to attract people who like to cook, and we have dinner parties about once a month. We cook things from different countries—we'll have an Italian dinner, a Greek dinner, or a Japanese dinner, and so there is a true family-oriented sort of caring atmosphere among the people in the lab. I'm sure that I set the tone for that because I enjoy it, but I also seem to attract people who like to participate in that sort of thing. The people who have been in my lab have continued to keep in touch years and years after they've left the laboratory.[2]

Science seems to place little value on things that make our journey through life more pleasant—comfortable surroundings, a cozy chat, helping a colleague—if they distract from the work. Modern science values efficiency, rapid generation of data, being first, critical review of ideas and theories, quick results . . . and progress. What could nurturing possibly have to do with science? At first glance it seems irrelevant, in another world apart from science. And that is precisely my point.

A nurturing attitude can balance the grimly focused efficiency that strips the pleasantness from life. Our fast-paced world puts such a premium on efficiency that some laboratories have become data factories, and the pressure to produce results drives some people to dishonesty. In addition to the few cases of fraud that have been documented, subtle deceptions abound as forms of passive resistance. For example, in one lab several people routinely arrived in the morning, turned on an empty centrifuge (to give their advisor the illusion that they were working), and went out for coffee.

Unfortunately, the sustaining role nurturing plays in life is rarely considered important enough for historians to write about or scientists to study. Just as nurturing has been devalued by society, it also has been dismissed in science—even at the cellular level—as not very interesting. For example, the function of glial cells has largely been ignored, since these "helper" cells were thought just to feed the nerve cells and clean up afterwards—the "little lady role," as psychopharmacologist Candace Pert calls it.[3] Although glial cells are ten times more numerous in the brain than neurons (nerve cells), they have been neglected in favor of studying the more active, exciting nerve cells.

This disdain of neuroscientists to study cells that play a mere nutritious role forestalled findings that glial cells participate in communication between the brain and the rest of the body. By moving back and forth between the brain and the body (there becoming macrophage monocytes, a type of white blood cell of the immune system), glial cells destroy the myth of the blood-brain barrier—a physiological reflection of the Western belief in the separation of mind and body. Interestingly, the number of glial cells per neuron increases as mammals ascend the phylogenetic scale from mice to humans.[4] Perhaps an intriguing study of Einstein's brain will nourish interest in glial cells: Marian Diamond, a neuroanatomist at the University of California at Berkeley, examined Einstein's brain, which had been preserved for scientific study. Compared to eleven other normal human male brains, she found that Einstein's had the most number of glial cells per neuron. The difference was particularly significant in the area associated with the conceptual powers of imagery and complex thinking.[5]

Because the process of nurturing is not dramatic, it is often ignored. However, it is not passive. It is intensely active, but has a different rhythm, a rhythm that is cyclical and repetitive—attending that continues hour after hour, day after day, year after year. In its constancy it often becomes invisible. Nurturing does not create instant rewards; it is a never-ending process. It requires patience and caring labor, connection to the object of care, empathy, compassion for the process of learning, gentle guidance, nourishing, and protecting. In our culture, women traditionally have carried the function of nurturing in that they have assumed the primary responsibility for caring for and teaching children, and nursing the sick. But it is also a function that men practice when they act as mentors.

NURTURING STUDENTS

The process of feeding the minds and souls of students is another vital aspect of nurturing in science. As nurturers, teachers are guides rather than authorities. They provide a safe space for development and expression of the unique person in their charge, offering the student the tools necessary to participate in the scientific community. Zoologist Ingrith Deyrup-Olsen, who has been training students in the classroom and the laboratory for over forty years, cares greatly about her teaching:

> I feel that talking to students is very worth my time! The women scientists I know tend to be more open and cordial towards students, more committed to the next generation. Sometimes we feel the commitment to students is more important than the commitment to our particular research. What's going to happen on the cutting edge of the science of the future, as embodied in these young scientists, is just as important as what we're doing at the cutting edge today. That isn't an aspect of generosity on my part. I see it as a long-time commitment in which my life is a very short component compared with the future of the science. But I know that quite a few of my colleagues don't feel that way.[6]

College students encounter the cold shoulder from science when they confront the do-or-die mentality of many science courses. Introductory science courses are often designed like initiation rites to eliminate the unworthy. Like the army, the objective is to separate the men from the boys. Surviving the course is a heroic achievement. Instead of motivating students by making science exciting

159

and attractive, science professors announce to freshmen, "This is where you get weeded out."

I encountered a similar sink-or-swim environment while doing my graduate research. My advisor rarely, if ever, took time from grant writing to teach me lab techniques. He expected me to learn them simply by reading the procedures in a journal article. To a student new in the field this is not a simple matter, since the authors usually make a number of assumptions, use jargon unknown to the beginner, and leave unexplained the fine details of the technique. It is like turning someone loose in the kitchen with a gourmet cookbook, but without the foggiest notion of how to sauté, fold, or separate egg yolks from the whites.

Nurturing, on the other hand, comes from a position of love. Most commonly we think of it in terms of caring for the young or the sick. It is not heroic, dramatic, or exciting, but a quiet, ongoing process, like wind carving stone. As a relationship of trust and connection, it is structured by hopes and expectations. It is an intrinsic part of life composed of small acts of focused attention. Nurturing requires receptivity. Just as the final shape of wind carved stone depends on the nature of the stone, so does the form of the scientist depend on the unique nature of the individual. Nurturing develops and exposes the glimmering potential that resides at the core of the student's being, polishing that potential until it shines with a light of its own.

The zoology course taught by Aimee Bakken provides an example of nurturing students while stimulating them intellectually and expecting high performance. Her course challenges her students; at the same time she supports them through it. Her efforts are rewarded when students tell her, "This is the hardest course I've ever taken, but I'm learning *so* much. Thank you."

Bakken also strives to provide a nurturing atmosphere in her lab. Her students and postdocs say she does it in such a way that it gives them a chance to develop, to work for themselves, and to become independent. Bakken believes that if someone likes what they are doing, then they will be self-motivated. She does not motivate by standing over them cracking the whip or watching the clock, but instead shows interest in her students' projects and provides gentle guidance. I envied the individual attention Bakken gives her students. She says:

When a new student comes into the lab, I ask them to set aside a few blocks of time each week because I want to show them how to do things. I want to be there to answer questions myself while they're getting started, and then once they've got those things down, they can come in whenever they want. I tell them there are other people in the lab they can ask for help, but I clearly let them know I want to teach them how to do things.[7]

Once Bakken has shown her student a technique, she explains that there are many different ways of doing things, and that different people in her laboratory may give them a slightly different way of doing it. She tells her student the reasons why she does things certain ways and asks them to learn it that way before starting to fiddle with the technique. She observes:

I think they like the fact that they're getting the personal attention. If they screw up an experiment, they come and talk to me about it. While we go through it, they know I'm trying to help them figure out what went wrong. When we find out what it is, and usually it's human error, I tell them a story about when I made that same mistake, so they get an idea that everybody has to make some mistakes as they go along.[8]

In this way Bakken encourages her students rather than humiliating them for making stupid mistakes. Earlier in life, she suffered from criticism that implied she was stupid because she did a stupid thing. She keeps her experience in mind when training others, and is concerned about how they feel. She strives to take the sting out of criticism and points out errors without personally attacking the student. She knows that it relieves much personal anxiety to say, "Oh, the column has run dry," rather than "You let the column run dry."

Despite the fact that management seminars have espoused these techniques for giving feedback to people (criticizing the behavior, not the person), few scientists receive training in management or give much thought to the psychology of relationships. Callous criticism seems to be the norm. A new technician in Bakken's lab who had made some serious blunders is still uncertain how to react to her style. When she asked about his progress, at first he looked frightened, expecting her to "jump down his throat." When she did not, he did not know quite what to do about it. While aware of his mistakes, Bakken also sees how much he is learning.

A graduate student rotating through Bakken's lab recalls the contrast between nurturing and nonnurturing labs she has worked in:

> In one lab my boss was very aware of everything I was doing, and how it was moving from one step to the next. It was great—when my face brightened up she knew immediately what I had found because she was aware of what I was doing. Her attitude was tremendously supportive and encouraging. Being in her lab I learned huge amounts of material.
>
> In the other lab the professor had a lot of administrative and teaching responsibilities, and so wasn't paying much attention to what I was doing. It was not a great learning experience. I didn't know if I was learning. I read papers and looked at the cells and saw different things, but I still to this day have no idea whether what I was perceiving as a significant observation was actually something you should be paying attention to. A beginner in science will look at something that a person who has been looking at a long time will be able to tell you "I've seen it before and it doesn't really matter that much, it's not that significant." I didn't get that kind of feedback, so things that I was chasing down may or may not have been real.

Bakken also is aware of her students and colleagues as whole people, not just as scientists. Her conversations weave together the personal and scientific realms. She tunes into her students' moods. If she notices someone dragging around or absent from the lab, she asks, "I haven't seen you in the lab much, are you really bogged down with course work, or what's happening? Are you stressed out? Feeling overworked?" Often she will preface her question with, "Look, I'm not trying to pry into your personal life, but the experiments are not coming along and I need to know whether it's because something has gone wrong with the method, or you don't have the time to put in now, or whether there is some way I can help you." She has found that students appreciate her concern and tell her openly what is happening in their lives.

In my twenty years in science I never felt comfortable talking to my professors or bosses about my personal problems, with the exception of a woman I reported to briefly. It was considered unprofessional. No one wanted to hear "excuses," no matter what was going on in my life, from pressures of exams to loss of a relationship or a death in the family. Nothing should interfere with the work. Personal problems must simply be ignored or overcome.

Only results mattered. But typically the science gets stalled until the personal dimension is addressed and problems are resolved.

Sensing that Bakken is accessible as a human being, students in her classes also instinctively seek her out when in trouble. Bakken recalls:

> I walked out of the lab one day and there was a student hovering outside my door. She had been in one of my classes, and I hadn't seen her for a few months. She looked like she was in pain in some way, so I asked, "Are you looking for me, Kathy?" She said, "Well, I don't know. I'm just sort of wandering around and I found myself here." So I invited her into my office to talk. She asked if I knew anything about melanoma. I said, "Yes, actually I do. I had it." Then she proceeded to tell me how she had just been diagnosed as having melanoma and she was scared shitless. We talked a lot to find out what her doctor had told her and what she needed to do then to get a second opinion. That kind of thing happens to me not infrequently.[9]

The fear of melanoma paralyzed Kathy. Until her conversation with Bakken, her fear blocked any productive work. By attending to Kathy's feelings and sharing her own experience, Bakken helped alleviate her fear, so that she could go forward with her life.

While such involvement in the personal lives of students and colleagues is not common in science, mentoring also occurs by way of introductions into the old boys' network, recommendations to review papers, and referrals to other labs. But sometimes it takes someone to "believe in" us—to recognize our talents and encourage us to reach beyond our own vision.

Biophysicist Cynthia Haggerty feels very fortunate to have had at least two superb mentors nurture her fascination with biology. Because no one in her family had ever gone to college, Haggerty had no such expectations until a high school biology teacher recognized her interest. She supplied the encouragement and recognition that Haggerty's family could not provide. She gave a focus to Haggerty's "puttering" by suggesting she enter the regional science fair, something it would never have occurred to Haggerty to do. Much to her surprise, Haggerty won a scholarship for a year in college for her project. Over the years, her teacher became a very good friend and they still keep in touch a couple times a year. One of her college biology professors also encouraged her work and coaxed to keep going with it. He later facilitated her entry into graduate school and is also still a friend. Without this personal attention and nurturing, Haggerty would never have tasted the excitement of scientific discovery.

NURTURING IDEAS

Ideas that go beyond the current scientific paradigm often meet with ridicule or swift dismissal by the critical and skeptical scientific community. Examples of this abound, such as Barbara Mc-Clintock's work on mobile plant genes and Rupert Sheldrake's hypothesis of formative causation (which we discussed in chapter 4). Many less courageous researchers abandon their work when they receive no support, aborting the life of the idea. When a concept invokes the derision of colleagues, such as when an editor called Sheldrake's book the "best candidate for burning there has been for many years," other scientists tend to avoid it rather than risk their reputations by becoming associated with "pseudo-science."

Criticism plays a valuable role in science. Out of fear of rejection or fear of looking foolish, criticism forces people to carefully think through their ideas and present convincing evidence to support them. Critical comments from colleagues compensate for individual blind spots or wishful thinking. Critics point out holes in the data and push scientists to hone their theories. But all too often, criticism comes too early in the process. It has become an automatic response, rather than a response modulated to the level of development of the idea. As a result, many ideas die stillborn. Here is how one biologist described criticism in the laboratory. "There is an enormous amount of talk in most laboratories. On the grounds that if you cannot persuade your colleagues you won't persuade the outside world, the rule in our lab is that you must be critical of your colleagues' ideas to the point of brutality or rudeness. That's the way to test them."[10]

This prevailing critical stance reflects the predominance in science of the male conversational style as described by Deborah Tannen. She and other linguists have documented that women ask

fewer and shorter questions. They are more likely to offer suggestions and ask questions to clarify or elicit more information (thus
admitting their ignorance). Men, on the other hand, use questions
to exhibit knowledge and negotiate status. They tend to preface
their questions with statements, ask multiple questions, and follow
up the speaker's answer with additional questions or comments.[11]

Exemplifying this masculine style, Robert Gallo at the National
Institutes of Health writes of his love for "the rough-and-tumble of
intellectual debate" and says that his weekly staff meetings can be
"grueling." In his book on the discovery of the AIDS virus, Gallo
describes the "merciless" questioning of scientists who must be
prepared to defend research they present at scientific meetings. He
quotes his colleague Genoveffa Franchini as saying,

> If you talk with any young postdoc about the Cold Spring Harbor
> meeting, they are terrified. In fact, Antoine Gessain, a postdoctoral
> Fellow from Lyon, France, told me a story about waiting for his turn to
> speak when this guy, another speaker, fainted. It is really very high
> pressure for young people. You're expected to go there, but not un-
> prepared. The interrogator can ask you whatever he wants. And that's
> always done. The questioning can get very nasty. But it's the best
> meeting for young people in biology, especially in human virology.[12]

While rigorous proof is the keystone of science, the seeds of new
ideas and theories need a safe and protected environment in which
to mature before confronting the harsh light of criticism. Propo-
nents of fragile, half-formed ideas often feel vulnerable. They hesitate to open themselves up to ridicule for suggesting a silly concept.
On the other hand, nonjudgmental brainstorming with supportive
colleagues helps ideas to blossom. By balancing skepticism and
criticism with experimental belief and a nurturing attitude, scientists can help their colleagues gestate new ideas. Then it takes time,
effort, and money to obtain the data to nourish and support new
ideas. Sadly, it is much easier to get funding to extend our knowledge about things we already know than it is to break new ground.

Cynthia Haggerty actively searches for colleagues to bounce
ideas against, to find people interested in her work who will listen
and brainstorm with her. Early experiences with teachers who
encouraged and supported her in high school and college taught her

the tremendous value of connecting with others to nurture ideas. Haggerty said:

> Early on, I found who the people were who would jump on me, so they just didn't hear from me anymore. But I kept looking until I found somebody who could sit and work with me and listen to me. I'd lay out my ideas and data and get people's opinions on them and they'd ask, "Well, have you looked at this or that?" I enjoyed that. It was a part of how I did my science and why we were so productive publicationwise, publishing about five papers a year. Over the years, particularly going to meetings, I found people I clicked with—people I really enjoyed, and who knew their work well. So I'd call up somebody halfway across the country and say, "I've got this idea and this piece of data, and thought of doing so and so. What do you think about that?" and then they'd answer. So I actively searched for people I could do that with, and talked about the weather to the other ones.[13]

In turn, colleagues from all over the country call to talk about their ideas or difficulties when they are stumped by puzzling data.

A nurturing approach involves maintaining an openness to many possibilities, different ways of developing. It allows unconscious processes, the movement of the spirit, and the voice of the soul to speak rather than limiting us to information from only rational sources. Entering this liminal, somewhat chaotic state allows for the negotiation of new meanings and connections. Seeking clarity, clear perception, or fixed ideas too early is like aborting the potential growth of a germinating seed by unearthing it too soon. Nurturing is interactive with the student, an idea, or with nature itself. It is a process—a process that is iterative and responds to feedback.

Bakken holds weekly lab meetings to give people in her lab an opportunity to present their data, receive feedback from their colleagues, and brainstorm on the possible implications of the results. Going around the circle, they describe their experiments and lay out the data they obtained. Bakken asks the researcher to interpret the results, and then asks other people in the group to comment before she offers her observations. She watches with pleasure as they gradually make sense of the data. She said, "Sometimes you see the light go on, and it's just wonderful." A graduate student working in Bakken's lab offered her view of the process of exploring ideas:

With Aimee it goes really well. One time I thought I had a very interesting idea for a result, her eyes just popped open and she thought about it and said, "That could be." It made sense. For the rest of it, the times when we were thinking theoretically about the science we were actually doing, the data we were actually getting, was usually trying to figure out why it looked so weird and what the problems were.

Other advisors are not as open to developing their students. A thirty-eight-year-old graduate student in microbiology observed:

As a graduate student it's difficult at first because you're so ingrained with the idea that professors know more than you do. But as a grad student you have the opportunity to read a lot about one subject and to become an expert on that subject. Then after reading enough and thinking enough, you come to the realization that *you* have ideas, and you start to defend them. At first that may be difficult because your main professors aren't used to seeing you in this light and you really have to fight for it.[14]

Those who work in Stephen Hawking's lab in the Department of Applied Mathematics and Theoretical Physics at Cambridge consider interaction and brainstorming so important that they structure it into their day. Twice a day, everyone gathers in the Interaction Room for tea and theory.

INSTITUTIONS

After six years in graduate school at Purdue University and four years in a postdoctoral position at Harvard Medical School, cell biologist Diane Horn obtained a tenure track position at the University of Southern California. She arrived to find that her lab had been the repository for all the departmental junk, and she had to evict squatters before she could make room for her lab equipment. She faced one obstacle after another with no support from the administration or her colleagues. Months went by before the bureaucratic system processed her orders for equipment. Even obtaining something as basic as deionized water to prepare tissue culture medium took months of persistent struggle. In addition to establishing her lab, she was assigned an unusually heavy teaching load her first semester. On top of that, she felt she could not say "no" to a colleague who asked her to be a guest lecturer in his course. At the time she did not want to appear uncooperative; she later realized that he routinely found others to lecture for his classes so that he could spend more time on his research. Alone in a new part of the country, she had no one to turn to for encouragement and support.[15]

Many young scientists feel isolated and beleaguered as they struggle to make a home for themselves in the world of science. Sometimes even equipping the lab with basic facilities requires heroic efforts. Becca Dickstein reported that, after arriving at the Department of Bioscience and Biotechnology at Drexel University in September 1990, it took over eight months of writing memos and demanding support to get the lab wired with sufficient electricity to accommodate her equipment. It took fifteen months to get the gas lines in the ceiling hooked up to the lab benches. Then the air conditioner broke. By May 1992, she was still struggling to get the new air conditioner functioning before warm spring and summer days shut down research on her temperatures sensitive biological system.[16]

In contrast, Kristina Katsaros, Professor of Atmospheric Sciences at the University of Washington, recalls how her interest in meteorology was nurtured early in her career. While in Seattle as an exchange student from Sweden, she walked by the meteorology building one day. On impulse she decided to go in and find out more about the field:

> The chairman was a very fine gentleman, father of four girls, who sat me down and invited me to try the field and see how I liked it. He made me feel at home. That first summer I took some courses and one of the old professors invited me to join a field project, which involved mountaineering, sending up balloons, and studying the wind around Mount Rainier. So that did it. That was such a wonderful experience! Meteorology was a challenge. It was never all that easy. I had to work at it. When I graduated, I slipped into a research position there. I had finished my degree and my professor was off on sabbatical and had a little money to finish some project he had started. He died while he was away and then another professor invited me to join his group for a while. I immediately wrote my own proposals and brought in my own money. That went on for a while, and then I asked to teach one summer, and I got into more things. I had small children in those years

and just worked part-time so I was kind of in the background, and I kept it that way. I was grateful they let me do my thing. It went quite well. I didn't have a lot of tension when I was raising my children, and I'm really very grateful for that. I could have asked for more challenges maybe if it had been a different time; this was the early seventies, and affirmative action and all the pushiness started in the late seventies. I felt sorry for some of the young women who had all the limelight—they had to perform, they had to make a career. I could be a mama and enjoy it. A male colleague who is a very accomplished guy had to give up some of that part of his life. Most men don't get to be so close to their children. He said to me once, "You have the best of both worlds."[17]

Listening to the stories of young assistant professors trying to establish themselves, it struck me how much support and nurturing could help at a time like this. Typically everyone is so caught up in their own research that the question of supporting others rarely occurs to them. Yet, it would mean a great deal if colleagues actively welcomed the newcomer and physically helped set up the lab, like helping a friend move to a new home, or pitching in to help a neighbor raise a barn. Someone could take a personal interest in getting them settled, teach them the ropes, and help them navigate through the bureaucracy. There are many ways science could incorporate the processes women have developed for networking and supporting each other's efforts and visions. In the current system, if someone doesn't make it, there are others ready to take their place. Paradoxically, while the scientific institutions are bemoaning the fact that there are a diminishing number of people entering science, they do little to encourage people to *stay* in science.

Today, when young scientists slam up against the problem of getting funding, they need to have the self-confidence to say "this is unjust" rather than "I am unworthy." Some need to feel that someone believes in them. Having someone with whom to share their frustrations would help prevent them from taking it personally, from being so hard on themselves and thinking "there is something wrong with me and they're doing this because I'm not important and don't deserve support." Instead of feeling embattled and spending energy just keeping their heads above water, they could be much happier and more productive if their colleagues were excited about them being there and encouraged them. Then they could put their energies into their research and have something left over to give to students and to invest in the personal side of their lives. In many areas of science today it takes an egoist to persevere. Katsaros observes:

> They lose an awful lot of good people who don't have the political verve, the get-up-and-go, the psychological support at home, or whatever it takes to fight all the battles to keep your research funded—to produce and manage everything. It's quite demanding. So many of my colleagues are not doing it anymore. They are doing something different. One colleague I collaborated with for years dropped out to take a trip around the country. Now, we don't know what he is doing. Another Ph.D.-level colleague of mine, a very capable guy, ended up working in computer sales. Another one, who got his Ph.D. in oceanography, is now doing analysis of data for the medical school. There's nothing wrong with these changes in principle, but I know the people involved are not themselves happy about it. All three are very bright and able, and should have been given more of a chance. These are people who have tremendous talent in logical analysis and all that, but they may not have the nerves or the communicative personalities, so we lose them. It seems there is something wrong there. I think that's where this feminine aspect comes in, a little more gentleness, more nurturing of the talents in people who are racehorses and who do need some freedom, a little comfort and consideration.[18]

After some twenty-five years of education (including four years of college, four to six years of graduate school, and one to four years of postdoctoral positions), an academic scientist finally obtains a university appointment and begins to set up their own lab. Again moving to a new place, usually after years of a fragmented lifestyle, the young assistant professor takes on new responsibilities of teaching, and serving on departmental and university committees, while writing grants, trying to attract graduate students, order equipment and supplies, and physically set up the lab. The pressures at this time are tremendous. Their performance during these first couple years sets the stage for the remainder of their career. If an assistant professor does not get funded for three years in a row, it's unlikely that they can remain at a leading university. Yet again the assistant professor confronts a sink-or-swim situation.

Institutions of science, such as the universities and funding agencies, suffer from a lack of nurturing. The system rewards scientists for the number of publications, presentations, and patents—not for teaching or participating in cooperative activities. Tenure guarantees professors a salary for life unless they commit a serious violation of the rules. To get tenure, a professor must publish enough research to impress a committee of superiors. Publish or perish. The quality of teaching is not considered to be important. The problem of getting tenure creates so much pressure on young professors that the universities cannot do the job they were built to do—teach students.

"The emphasis on research is certainly not in the best interests of the students," says Jack Gilroy, former dean of education at Seattle University.[19] In many of the major universities people get so engrossed in their research they do not have time to teach. At some large universities, more than 50 percent of the courses are now taught by adjuncts and graduate students. Some are good; few are around for very long. Good teaching requires creativity and continuity. In addition, more and more graduate students teaching courses are foreign and their English can be difficult for students to understand.

Awarding tenure and promotions based on the number of publications creates a system based on quantity rather than quality of research, and encourages publication of the minimum publishable unit. This has resulted in an overwhelming 74,000 science and social science journals, only 4,500 of which are indexed by the Institute for Scientific Information. Within those journals indexed, 55 percent of the papers were never cited in the five years after they were published. Allen Bard, editor of the *Journal of the American Chemical Society*, observed, "In many ways, publication no longer represents a way of communication with your scientific peers, but a way to enhance your status and accumulate points for promotion and grants."[20]

A reward system based on quantity rather than quality discourages long-term studies or the synthesis of several lines of investigation into a larger picture. It results in fragmented presentation of data and research programs. It fuels the drive toward speed and efficiency in order to be first, no matter what, whatever it takes. The end justifies the means. Emphasis on such goals can remove the humanity from an endeavor. In such a system one cannot be vulnerable.

To counteract this proliferation of papers, some institutions are now beginning to limit the number of papers they will accept for evaluation. The National Science Foundation (NSF) seeks to emphasize the quality of publications by permitting researchers to submit no more than five publications with their grant applications. Some elite universities such as the Harvard Medical School will review only five to ten of the applicant's most significant papers.

When it comes time to present their results at professional meetings, scientists again struggle for status, with little room for nurturing. Most conferences are organized as a tightly scheduled series of individual ten-to-fifteen-minute presentations, where one person presents their work to a roomful of people. Everyone sits face forward and listens to the person at the podium. Little or no time is scheduled for interaction. Frequently, the talks extend into the brief three-to-five-minute question period. While poster sessions (where researchers post their data on bulletin boards and discuss their results with interested scientists who stop by the exhibit) are more interactive, they are not considered to be as prestigious as giving a talk. Although the talk may summarize work produced by a team of researchers, the presenter rarely defers to a colleague in the room to answer a question. No one dares to admit they don't know it all. Even a small meeting of sixty researchers of the American Geophysical Union followed the pattern set by larger meetings. A graduate student in oceanography observed:

> The majority of people in the audience had been to sea with each other for weeks, working together under difficult circumstances. I thought that because we all knew each other it'd be all open and less formalized. But I was shocked that they had to dress up for each other in jackets and ties—and all the posturing and filibustering, like they have to make their own statement by being critical of another's work. Scientific discussions don't have to be confrontive.

Several women commented on the zeal with which male researchers "stomp on people," including inexperienced and junior people who are no threat. While some scientists ask questions to clarify a point, many scientists ask questions or point out holes in the presenter's data to show how sharp they are. In retaliation, some speakers try to make the questioner feel like a fool. Meanwhile, many graduate students and postdocs yearn for a forum at scientific meetings where they can present their work-in-progress to get feedback—where they can throw out data for suggestions, brainstorm interpretations, and interact in a nonthreatening way.

In an effort to create a forum to foster interdisciplinary exchange of ideas, a "working group" of seventeen women gathered in Santa Cruz, California, to discuss integrating the role of female biology into evolutionary theory. Instead of the traditional approach of presenting data, they focused on interaction and synthesis. They called their invitational conference "Women Scientists Look at Evolution: Female Biology and Life History." Over several days of intense, focused interaction, they devoted themselves to assessing, integrating, and developing theory. There was a noticeable contrast to the contentious adversarial atmosphere common at most conferences. Physical anthropologist Silvana Borgognini Tarli of the University of Pisa observed, "The atmosphere was almost perfect, the sort of atmosphere that should be present at all conferences, which are, after all, for communication. No one was searching in others' work for feeble points to attack. We had discussion without victory or defeat."[21] One of the themes of the conference was self-reflection: the participants discussed how they came to ask particular scientific questions and how their personal life histories affected their research perspective. In order to improve communication between scientists and the public, this conference uniquely included a public forum and active participation of science writers. Unfortunately, the article in *Science* covering this conference was accompanied by an insulting cartoon of little girls playing at science in a treehouse, sticking their tongues out at the football players below.[22] Although the all-woman format was a side effect of the initial selection process,[23] several letters to the editor accused the organizers of this "appalling" conference of sexism and "breathtaking audacity."[24]

THE LONG-TERM APPROACH

Due to the long-term responsibilities of gestating and raising children, a generational perspective has become attached to the Feminine. We are just now realizing the hidden costs of the short-term

173

approach to science and technology development—such as safe disposal of toxic chemical and nuclear wastes, soil depletion, and the squandering of natural resources. In looking for short-term payoffs, we rarely consider the health and prosperity of future generations. Although the nurturing approach is fundamentally process-oriented, it is always with an eye to the long-term results and consequences.

Modern science in America is basically an impatient field. Researchers rarely have the patience for the questions that take a long time and require a lot of thinking and listening and not just doing. In the push to get fast answers, geneticists turned from corn to fruit flies, then to bacteria, and finally to viruses as model systems for reproduction. Compared to the annual cycle of corn, fruit flies have the advantage of yielding a new generation every fourteen days; bacteria divide every twenty minutes, and bacterial viruses (phages) replicate several times in ten minutes. Meanwhile, the patient approach of the naturalist making observations over long periods of time became more rare. Still studying corn in the 1980s, Barbara McClintock seemed an anachronism. As a cell biologist, Aimee Bakken finds that working with whole cells, rather than molecules in a test tube, requires more patience. "Biochemists I've known find it too slow to work with cells. They want answers right away. They want answers every day. That's never really mattered to me much. I have patience to wait and see."[25]

Some people who come to work in Bakken's laboratory lack the patience and want quick answers. Uncomfortable with the slower pace of cells, they may leave her lab and seek model systems that promise quick answers. In addition, the competitive bidding for two- or three-year grants keeps research focused on the short-term. Bakken tries to combine the short-term and long-term perspectives. She tends to tackle hard problems that others avoid, and then stick with them until she comes up with interesting answers. Her choice of words reveals her nurturing approach to experiments: "You have to believe in what you are working on to make it work. It's not a matter of trying to force specific answers on it, but the answers won't reveal themselves unless you believe in it and keep working at it."[26]

While she recognizes the importance of publishing and sharing the results of her work, Bakken loves the process itself more than the results. When she spends too much time at her desk, she gets unhappy and frustrated. She would much rather be in the lab doing the work and looking at cells. Although positive results come slowly, the work gives her long-term satisfaction. Like the process of solving puzzles, she loves to "take all the pieces and put them together and see what it's trying to tell you."

Most field biologists spend a couple of months in the field collecting data and then return home to write up their results. Long-term studies that span decades are rare. Before Jane Goodall, Henry Nissen went to French Guinea to study chimps for two-and-a-half months in the 1930s. In contrast, Goodall studied generations of chimpanzees over three decades, following the lives of many individuals from birth to death. Before Dian Fossey devoted eighteen years to studying the mountain gorilla, the longest study of these animals was conducted by George Schaller during a year in Zaire in the 1950s. While Biruté Galdikas has been among the orangutans for over eighteen years, the longest previous study was undertaken by John MacKinnon and lasted three years. In the spirit of adventurers moving onto the next conquest, each man went on to study other animals in other places—"the typical way males do things," as Galdikas remarked.[27] Louis Leakey believed that women were more tenacious than men and particularly well suited to long-term studies. Each of the women sponsored by Leakey nurtured her relationship with her chosen animals, gaining their trust by remaining harmless and receptive while observing them. Beyond their scientific work, these women devoted themselves to tending the sick and injured, working to protect the apes from poachers and to preserve their habitats, and lobbying to improve conditions for animals used in scientific research.

Like any quality of the Masculine or the Feminine, nurturing also has its dark side. Just as the masculine approach of efficiency—the one-sided extreme of getting the job done no matter what—strips the pleasantness from life, so does a one-sided feminine approach have its perils. Like overly protective "mothering," too much nurturing stifles creativity and innovation. Carried too far, it can interfere with the science. Because of the scarcity of nurturing in science, those who do nurture can be overburdened with students seeking mentoring or troubled colleagues who need to talk—to the point that it interferes with the nurturer's own productivity. If boundaries are not set, a few people can end up carrying the feeling function for a whole department. For example, when it comes time for departmental functions such as the annual Christmas party, Bakken is asked to organize it much more often than the male faculty members.

Science is a process of learning and discovery—and we are vulnerable while we are learning. In order to freely question and explore, we must accept vulnerability in ourselves and others. Without compassion and respect for our vulnerabilities, we cannot have a learning community. A more nurturing attitude in science allows us to expose our ignorance without fear. Let us learn to value sustainability equally with progress, and recognize the importance of collective small acts equally with impressive heroic deeds.

Next, we will explore how a new view of nature appears through the eyes of the Feminine, and how corrosive competition diminishes science.

COOPERATION
Working in Harmony

COOPERATION IN NATURE

When he first presented his theory of evolution to the Linnean Society, Charles Darwin opened his paper with an image of nature as a brutal struggle between opposing forces: "All nature is at war, one organism with another, or with external nature."[1] At the same meeting, Alfred R. Wallace, the codiscoverer of natural selection, described plants and animals engaged in "a struggle for existence, in which the weakest and least perfectly organized must always succumb."[2] In *The Origin of Species*, Darwin states that in "the universal struggle for life . . . all organic beings are exposed to severe competition."[3] (Interestingly, Darwin has been accused of stealing his ideas from Wallace.) This view of nature as inherently competitive continues to dominate the perspective of Western science.

However, there is emerging a cooperative worldview that can stir us to see new connections in nature and lead scientists to ask different questions. In their book *The New Biology: Discovering the Wisdom in Nature*, Robert Augros and George Stanciu[4] explore examples from the fields of biology, paleontology, and physics that portray nature itself as "a cooperative alliance rather than a cruel and indifferent hunting ground" that awards survival only to the fittest individual. Augros and Stanciu point out that nature avoids competition by dividing the habitat into ecological niches. Rather than compete for the same food or shelter, species adapt to specific types of food, feeding times, or living conditions. For example, three species of yellow weaver bird in Central Africa live peacefully side by side. They do not compete over food, since one feeds on hard black seeds, the second prefers soft green seeds, and the other eats only insects.[5] Animals such as cats mark their hunting ground with scent marks to guard against undesirable encounters. Similarly, researchers stake out territory: "I work on this aspect (field, level, organism) and you work on that aspect." For some, communication of results to other researchers is more a strategy for staking out territory than a matter of sharing information for the purpose of discussion. In doing so, they reduce the possibility of duplication of effort.

Another of nature's strategies for peaceful coexistence is voluntary restraint. This flies in the face of the Darwinian model, which maintains that a species will increase without limit unless checked by predation, starvation, disease, or severe climatic change. However, field studies demonstrate that many animals limit population growth by benign internal methods, such as ovulating more slowly or stopping ovulation altogether, as in the case with mice in overcrowded conditions. Animals such as elk, bison, moose, lions, sperm whales, and harp seals defer the age of sexual maturity whenever overcrowding occurs.[6] Other animals, such as the white rhinoceros, rely on celibacy where herds contain a large number of nonbreeding subordinate adults.[7]

In modern agriculture, weeds are considered to compete with the crop plant and so are eliminated by herbicides. Scientists at the University of California at Santa Cruz recently studied the traditional practices of Mexican farmers who prune back, rather than pull, a weed that commonly sprouts between rows of corn. Researchers found that the roots of the weed, *Bidens pilosa*, secrete compounds lethal to fungi and nematodes that destroy corn. The farmers simply trim the weed fifteen days after the corn emerges and every thirty days after that. Instead of competing with the corn, the weed controls the pests without significantly stealing soil nutrients from the corn.[8]

Plant physiologist Frits Went observes: "There is no violent struggle between plants, no warlike mutual killing, but a harmonious development on a share-and-share basis. The cooperative principle is stronger than the competitive one."9 Furthermore, according to Went, weeds threaten to crowd out crops only when the crops have been planted out of season or in the wrong climate. Even in harsh environments such as the desert, or habitats with a profusion of plant life like the jungle, Went finds peaceful coexistence rather than one species crowding out the other: "In the desert, where want and hunger for water are the normal burden of all plants, we find no fierce competition for existence, with the strong crowding out the weak. On the contrary, the available possessions—space, light, water, and food—are shared and shared alike by all. If there is not enough for all to grow tall and strong, then all remain smaller. This factual picture is very different from the time-honored notion that nature's way is cutthroat competition among individuals."10

Because of the assumption that competition is the way of nature, research into its cooperative aspects receives little attention—or funding. Biologist William Hamilton observes, "Cooperation per se has received comparatively little attention from biologists."11 Zoologist Robert May notes that the topic of mutually beneficial association between different kinds of organisms has remained relatively neglected in field, laboratory, theory, and textbooks.12 In reality, organisms help each other in a variety of ways. One organism can use another as a protected place to stay, such as certain crabs that live in the rectums of sea urchins. Many ants live in mutually beneficial association with particular trees or shrubs. For example, the ant *Pseudomyrmex* lives in the hollow thorns of the *Acacia* tree of Central America and feeds on the nectar on the leaves. To keep its *Acacia* home healthy and ensure a continuous food supply, the ant drives away plant-eating insects and prunes back vines that threaten to crowd out the *Acacia*.13 Other insects help plants achieve cross-fertilization in exchange for nutritious nectar. Many species provide transport for other organisms or its seeds. Others live by cleaning. The Egyptian plover eats parasitic leeches inside the mouth of the crocodile and emerges safe and well fed. Blowflies lay their eggs in festering wounds. When the larvae hatch, they feed on the pus, consume dead tissue, and disinfect the wound with their excretions. Such relationships illustrate the value of community and interdependence as opposed to fierce individualism.

At the most fundamental level, life is a cooperative enterprise: Plants absorb carbon dioxide from the air, energy from the sun, and water from the soil to produce sugars. They release oxygen as a by-product of photosynthesis. Animals then breath in the oxygen and eat the plant sugars, breathing out carbon dioxide and excreting water (urine) in the process. Neither plants nor animals can exist without each other—or without the plethora of bacteria that decompose organic matter into usable components.

SYMBIOSIS, THE SCIENCE OF COOPERATION

The actual study of cooperation in nature is best exemplified by symbiosis, the subfield of biology that studies unlike organisms living cooperatively together—such as the complex association of fungi and green algae we call lichens, or the bacteria living in a cow's gut that help digest cellulose. In a sense, symbiosis is a uniquely feminine topic of research. As a science of relationship, it provides a model of interdependence at the biological level.

In contrast to well-funded research on disease-associated parasitism, scant funding is allocated to studying healthy associations of organisms. Ironically, the most successful parasite is a symbiont, because it does not kill its host and thereby lose its home. By neglecting the study of symbiosis we do not consider possible benefits that a parasite might eventually bring to its host. Fixation on a competitive model can blind us to seeing the cooperative side of nature.

The work of Lynn Margulis illustrates how the Feminine shifts our perspective and gives rise to radically new questions about nature. Instead of assuming that competition is a law of nature, she explored the role of cooperation in evolution. She writes, "Although they are often treated in the biological literature as exotic, symbiotic relationships abound; many of them affect entire ecosystems."[14] Margulis presents convincing evidence that biological diversity arose as much by microbial cooperation as through Darwinian competition. Twenty years ago she postulated that nucleated cells (such as the cells in our bodies) evolved from more than one type of bacteria—that nucleated cells are bacterial communities that coevolved. In other words, one bacteria ate but did not digest the other, and now they need each other to survive. At first, her theory of cell origins was considered "offensive" and "scandalous" and could not be discussed at respectable scientific meetings.

Margulis has written over 130 scientific articles and seven books. In 1983, she was elected to the National Academy of Sciences for her work, for an American, an honor second only to winning the Nobel Prize. Now almost all biologists agree that mitochondria (organelles in the cell that produce energy) were once oxygen-producing respiring bacteria, and that chloroplasts (organelles in plant cells that photosynthesize) were originally cyanobacteria. Margulis insists that such changes could not have come from an accumulation of chance mutations. In fact, she maintains that "the major source of evolutionary novelty is the acquisition of symbionts—the whole thing then edited by natural selection. It is never just the accumulation of mutations."[15]

Most theories of evolution emphasize mutation as the major source of new genetic information. In Margulis's arguments, however, symbiosis is far more innovative in generating biological novelty than is the accumulation of chance mutations, although the latter is more commonly credited as the basis of evolutionary change. Mechanisms of mutation such as base-pair changes, deletions, duplications, and transpositions derive from a reductionistic perspective. Alternative mechanisms, such as an increase in the number of sets of chromosomes, derive from a more synthetic perspective—but have been largely ignored. Now, however, some interest is picking up. Two other women, Lynda Goff of the University of California at Santa Cruz and Annette Coleman of Brown University, are studying the acquisition of foreign nuclei as a mechanism of evolution. Neil Todd at Boston University and a handful of other researchers are also presenting evidence for various synthetic mechanisms.[16]

In spite of the acceptance of her theories, Margulis observes that, "Most of my colleagues would agree that mention of symbiosis in a grant application tends to deny funding."[17] Symbiosis remains an obscure, virtually unfunded subfield of biology. It is either ignored or merely defined in the major textbooks on evolution. In 1983, however, the International Society for Endocytobiologists (ISE) was formed to provide a forum for researchers studying intracellular symbionts. Now two journals founded in 1984 and 1985, *Endocytobiosis and Cell Research* (published by ISE) and *Symbiosis*, unite biologists from disparate fields. Over the years, Margulis's work has captured the imagination of many lay people as well as scientists. Even those who disagree with her acknowledge that she has stimulated their thinking by offering them a new perspective.

Other researchers are beginning to look at the practical applications of symbiosis. For example, agricultural funds have supported studies of nitrogen-fixing bacteria that live in root nodules of legumes such as clover, peas, beans, and alfalfa. Some molecular biologists hope to add nitrogen-fixing genes to other crop plants. With the ability to fix nitrogen, there would be less need for fertilizers.

In addition to its practical applications, the symbiosis between plants and bacteria can be a way to listen to the conversation that turns genes on and off in the process of development. As molecular biologist Becca Dickstein says:

> In some ways it's like a dance of development. Both of these guys are making music, and they're playing together. Some of the notes have been identified from the bacterial partner, and I have a glimmer of a clue as to what the notes are from the plant partner. I want to go in and find out what those notes are, and find out what the next step is! Then we'll find out if there is a crescendo or whatever. There is this back and forth, from the bacteria to the plant, from the plant to the bacteria. I want to know what it is and how it works! And at what point there are big things said and at what point there are little squeaks, and what those big booms and little squeaks mean in coordinating the development of this organ [root nodule] that can't be made without these contributions from one side and then the other side.[18]

Cooperation in the form of symbiosis is not just a rarely occurring quirk of nature, but a fundamental force in survival and adaptation. Symbiosis research shows that, although competition plays a role, it is not the dominant law of nature. In fact, the field of symbiosis itself is particularly cooperative—according to Dickstein—perhaps because the research itself centers around organisms working together.

Taking a symbiotic approach can even change the way we treat disease. Who would have thought that healing cancer cells would heal the patient? Yet, when faced with the rare but difficult-to-treat acute promyelocytic leukemia (APL), scientists were forced to seek a new approach because the chemotherapeutic drugs exacerbate the excessive bleeding caused by the cancer. As opposed to the "war on cancer," killing all cancer cells with radiation and chemotherapy, researchers tried using drug treatments to help malignant cells grow normally to maturity. Several groups now report cures of certain cancers using retinoic acids, compounds derived from vitamin A. In a clinical trial conducted in China, a retinoic acid cured all of twenty-four APL patients; in a French study, fourteen of twenty-two APL patients were cured. Researchers at the Memorial Sloan-Kettering Cancer Center in New York City report a retinoic acid caused complete remission in nine out of eleven APL patients. APL, a leukemia that annually strikes 1,500 people in the United States, is characterized by the uncontrolled growth of a subclass of immature white blood cells bearing a specific chromosomal abnormality. As long as these cells remain immature, they are immortal. But the retinoic acids help these immature leukemic white cells to finish developing. Then the cells die naturally after a normal life span, and are replaced by young cells capable of normal development. Researchers also report success using this technique with advanced squamous cell carcinoma of the skin. This research leads oncologist Razelle Kurzrock at the M. D. Anderson Cancer Center in Houston to say, "It makes you think that we'll be able to solve cancer."[19] These cures come from taking the perspective of helping the cancer cell, rather than fighting it.

THE COMPETITIVE NATURE OF SCIENCE IN THE UNITED STATES

Competition drives much of science in the United States from the top down: the U.S. government supports basic research to assure the nation's future economic competitiveness and national security. Industry funds basic science in order to compete in national and international markets. Sigma Xi, the honor society of research scientists, affirms "the overall need for American competitiveness remains an important requirement, and the role of science and technology in maintaining competitiveness is vital."[20] In fact, up until 1984, antitrust laws forbid companies from cooperating for fear of collusion. It was illegal to share research.

An adversarial approach even drives the study of nature, as reflected in the language of scientists. Francis Bacon wanted to "conquer and subdue her [nature], to shake her to her foundations, to storm and occupy her castles and strongholds." Modern science talks about the "war on cancer." In contrast to this language of conflict, developmental biologist Aimee Bakken likens her approach to science to "opening up a flower to see what is inside."

The reward system in science fosters competition. Cooperative or nurturing activities such as committee work, providing research materials to colleagues, community service, teaching, and consulting are usually ignored when it comes to raises, promotions, or awarding grants. Honors or awards rarely recognize teams or groups. The Nobel prize, the most prestigious award in science, cites no more than three individuals per category. Daniel Koshland, the editor of *Science*, writes: "The unpredictability of success and the great effort induced by competition are essential for the rapid progress in science we all commend."[21]

The requirement for survival in science is publishing. In order to force scientists in the seventeenth century to divulge their data, the Royal Society of London ruled that priority goes to whoever publishes first, rather than who discovers first. Since then, researchers have scrambled to be first to publish. Little glory goes to also-rans or investigators who verify another scientist's findings— even though independent verification is an important tenet of science. In addition, scientific journals compete to publish ground-breaking research. While *Nature* and *Science* have amicably competed for years, their editors are worried by *Cell*'s attempts to overtake their papers. In one instance, *Cell* published a paper in fifteen days (compared to their usual four-and-a-half months) in order to beat out a paper by a rival team that was soon to come out in *Nature*. Or university administrators, eager to pull in more grant funding, pressure researchers to prematurely reveal their results at a press conference. This happened at the University of

Utah when B. Stanley Pons and Martin Fleischmann announced they had achieved "cold fusion."

Neurobiologist Richard Aldrich of Stanford University comments about the fierce competition in sequencing significant genes in the area of molecular biology: "Before, science was more interpretive. It was not so much to be first but to be best. But there is no such thing as a better sequence."[22] However, in the rush to publish, even these sequencing papers often suffer from incompleteness and minor sequencing errors passed over in the frenzied review process by fax. Graduate students, postdocs, and small labs often become justifiably paranoid about being scooped by big research labs that can put a team of experts onto a "hot" project. In a disturbing number of documented cases, prominent scientists, acting as reviewers, advised journal editors against publishing a submitted article or advised a granting agency against funding a research proposal, then quickly carried out the same project in their own labs and published it under their own names.[23]

Even within an organization, some research directors foster competition. For example, one researcher worked in a biotechnology company where the scientific director pitted one group against the other. He told each of them, "So-and-so is working on that, but I don't think he's going to get it," and then rewarded only the first investigator to obtain positive results. As a result, scientists withheld data and materials from each other out of fear of being scooped—from within their own group! They felt constrained from talking about their new insights and discoveries. Instead of pooling their resources and making large batches of materials to share, they constantly bickered over who had the right to use the limited materials. The researcher found the most difficult part of her work was "when people are very grasping about the work they've done and don't share—that's harder to accept than when equipment breaks down." Every time she started an experiment, she found someone had "borrowed" her reagents or equipment. She then wasted time and energy tracking them down. By nature she found collaborating with other people the most rewarding part of her work. She loved "sharing of ideas and the stimulation that one person gives to another, because the whole is greater than the sum of its parts." Sadly, her boss penalized her for doing "too much collaborative work" by giving her a poor raise and taking away one of her technicians. She then became paranoid about being cooperative. This environment slowly wore her down and eroded her desire to do science. Such examples may seem small and petty, but they reflect the daily impact of corrosive competition. This very dailiness makes it important.

In the past, science tended to be a pursuit or hobby of those with independent financial means. Well-to-do men such as Charles Darwin could follow their curiosity about nature without fear of losing their income. The monk Gregor Mendel could seek the truth of genetics on a small plot of land. Today, almost all researchers pursue science as a career. Their living depends upon it. Simply to be able to do science in the United States involves competition, since one must have resources—a position, funding, equipment, and usually technicians, assistants, and graduate students. These resources are limited and the cost of most twentieth-century science has moved beyond the reach of amateurs. James Watson, who won the Nobel prize for explaining the structure of DNA, noted that "not only the self-respect but also the way of life of a scientist depends upon his relative ranking among his peers."[24]

The fierce competition of modern science is paradoxical since the nature of scientific research depends on the basic assumption of mutual dependence and sharing of information. Each discovery is built on the foundation of knowledge and technology of others. In a section labeled "Cooperation, not competition," Sigma Xi states, "The mutual dependence of one scientist on the work of others, and the integrity with which this work is reported and published, are essential characteristics of the whole scientific enterprise."[25] Scientific meetings and scientific journals provide formal structures for this exchange of information. Despite this ideal of cooperation, the hierarchical structure of science fosters competition. While Paul Berg magnanimously talks about his research building on the work of the postdocs and students in his lab, he alone captured the glory of the 1980 Nobel Prize in Chemistry. He said:

> When the head of a team like mine gets recognition, everybody in the lab knows that the credit is shared among many. I have not done any research work with my hands for five years. The work may have been

initiated or guided by me, but much of the work has been done by students who are building their careers.

We work as a unit. Very often we cannot identify the origin of an idea. It will have been adapted and modified and changed so often, then it leads to something else, and finally there is a breakthrough.[26]

This system places students and postdocs in a double bind. The official description of group work as cooperative veils a prevalent disdain for hard work done well. Sticking to an assigned task out of loyalty and commitment to the group can cost young researchers their career. In order to gain status in the hierarchy, they must undertake independent, risky work. While told that they are building their careers by working "as a unit" in a lab like Berg's, they cannot "make a name" for themselves or "make it big" as a member of a group. Because of the ideal of cooperation, many scientists drop out of science when confronted with the reality of competition. Or they find they cannot simply do science for the love of science—and survive.

Janet Thomas, now a well-established neurobiologist and tenured professor at a major university, found herself in a psychiatric hospital before she realized how adversely the competitive environment had affected her—and her colleagues.

Throughout all this there was a great deal of support for me from people in my department whom I barely knew. I thought everything that had happened was my fault. That I was the only one who was depressed about our work environment. But it turned out everyone was feeling that way, but no one felt they could talk about it; they all thought they had to walk around acting like everything was terrific.[27]

Thomas entered therapy and joined a woman's support group that includes administrative assistants, postdocs, students, and professors. Her lab has become a place where people can have emotions and safely discuss them, where they can succeed by being real without suffering as Thomas did.

Like the conspiracy of silence that keeps people from revealing family secrets, members of the scientific community rarely reveal the dark side of science. Rather than find fault with the system, people think they have to keep their problems to themselves and obey a code of silence. Whistle-blowers are shunned and complainers do not get ahead. When someone leaves science, those remaining dismiss them as failures, rather than reevaluate the toxic system that forced the person to leave.

Not only women suffer from the competitive environment of modern science. Men do as well. An Asian postdoc educated in the U.S. who has left high-energy physics compared his career to others who made it. He believes he failed because he did not adopt the preferred aggressive style and did not initiate an independent project outside of his assigned work.[28] Atmospheric scientist Kristina Katsaros is saddened by the number of capable male and female colleagues who have left science because of the competitive environment, weary of fighting all the battles to keep their research funded. She reflects,

> There is something harsh about the whole competitive environment. Not everyone can handle that. [Heavy sigh.] I just put up with it, I guess. I compete, somewhat, but I try not to. I make it my philosophy not to get trapped. I try to be generous, to share. There are people who keep secrets from each other because they're so afraid someone will scoop them. I try not to do that kind of thing. I find I always get more by talking and listening than the other way around.[29]

Next, let us look at the dynamics underlying different types of competition and cooperation, and then explore how carriers of the Feminine weave cooperation into the fabric of their science.

THE DYNAMICS UNDERLYING COMPETITION

Engaging in the world from a presumption of hierarchical order, where each person continually jockeys for status, leads to a view of life as a contest—in Darwinian terms, the survival of the fittest. At their best, these contests provide an arena in which to compare a diversity of approaches. Competition promotes vigor by weeding out weak organisms or bad ideas. Ecologically, it pushes organisms toward exquisite adaptation to the environment. Economically, it spurs companies toward efficient production and to provide products and services that benefit the customer. In sports, it propels athletes to extend the limits of human potential. Clearly, competition at its best can be beneficial.

What turns such joyful contest into cutthroat competition? What prompts athletes to take hormones or scientists to fabricate data? Are competition and cooperation mutually exclusive?

As I thought about these questions, I came to the conclusion that we do not have enough words to describe different types of competition—and that the underlying motives are perhaps more important. I found a clue to these motives in Jung's observation that love and power (in the sense of power over someone or something) are mutually exclusive. As we discussed in chapter 3, when love is the underlying dynamic, power does not enter into it. Conversely, when the motive is power-over, love has no role. From this perspective, competition is healthy when energized by love of the work itself—rather than as a vehicle to power in forms such as profit, position, or status.

When I managed the product development of a small biotechnology company, I straddled the worlds of basic science and the business of making products for profit. The company struggled with issues of funding basic research that might lead to future products, versus immediately developing products so that the company would *have* a future. While I understood that holding research as proprietary knowledge gave the company a competitive edge, it conflicted with the researchers' desire to publish and inhibited the vital exchange of ideas and materials with outside collaborators.

As a start-up company of young scientists, a spirit of adventure infused us. The joy of creativity and the satisfaction of working together with competent colleagues toward a common goal inspired many of us to work heroic hours. To me, this was an example of healthy competition and cooperation at its best: we felt challenged to use our knowledge and ingenuity to make products that could significantly improve health care. Although our products ultimately faced competition in the market, a love of doing the science motivated our work. We cooperated in the same way a smoothly functioning sports team relies on the contributions of each member.

Healthy competition based on love can inspire creativity, augment motivation, foster diversity and flexibility, stimulate a striving for excellence, and push people to go beyond themselves. Biophysicist Cynthia Haggerty observed the dynamics in her lab,

"In an atmosphere of cooperation, fun and enjoyment, creativity can be expressed in a safe and happy environment. If people are competing in a nasty way, cutting each other up, it slows down the work. Competition can be healthy when it's in a teasing and jovial way."[30] The joy of discovery, the thrill of unveiling a new aspect of nature, and the love of doing science are often sufficient reward for some scientists—as we discussed in depth in chapter 3.

Competition crosses over to a power motive and becomes corrosive when it undermines cooperation and interferes with another person's capability to do good science.[31] While it spurs an invigorating diversity of approaches to problems, competition based on power-over motives such as ambition leads to waste from duplication of effort—or to errors, hasty publication of insufficiently developed research, or fraud in the rush to publish.[32] In addition to the more obvious waste of time, effort, and materials, the drain of psychic energy that otherwise could be applied to creativity is enormous. After a while, many become trapped in the habit of competition. Always looking for the next fight, they see only the enemy, sometimes even after the enemy has "surrendered." To conserve energy, they close off awareness of life beyond conflict.

Power-over competition can be a response to an underlying fear of not being good enough, resulting in the need to obtain validation by winning. The need to win "no matter what," to beat out somebody else—in order to be approved of, to prove oneself, to be noticed—diminishes science. To compete well under these conditions, scientists must hide their personal side, closing themselves off from the compassion of others. This nasty side of competition betrays trust and selects against vulnerable and open human beings. It prevents researchers from seeking help, and from generously sharing their ideas and work.

THE DYNAMICS OF COOPERATION

Riane Eisler's book *The Chalice and the Blade* describes research on matrilineal European societies from 7000 to 3500 B.C. that were based on a cooperative partnership model of *power-with*, rather than the dominator model of *power-over*. In these societies neither half of humanity ranked over the other and "diversity [was] not equated with either inferiority or superiority."[33] An attitude of linking, rather than ranking, prevailed. In our culture, boys learn competition and girls learn cooperation at an early age. Stereotypically, girls play house or play with dolls while boys are out competing in sports. Traditionally, women have clustered together in circles for sewing, quilting, pickling, canning, washing clothes, and watching their children. As we discussed in chapter 6, the circle, a symbol of the Feminine, also symbolizes the cooperative meeting of equals. Carriers of the Feminine prefer being related, helpful, and connected. They give others support in times of need and celebrate each other's successes.

Research by Matina Horner[34] and Phillip Shaver[35] indicates women learn more easily in a cooperative rather than competitive environment. While male students may thrive on competing, females prefer and perform better in situations where everyone wins. In her book, *Female-Friendly Science: Applying Women's Studies Methods and Theories to Attract Students*,[36] Sue Rosser suggests emphasizing cooperative classroom and laboratory methods in order to make science more attractive to females.

In our hierarchical culture, cooperation is often misunderstood to mean obediently following someone else's lead. However, with an appreciation for the value of diversity, cooperation becomes an active give-and-take among equals. Fueled by a spirit of generosity, cooperation engenders creativity through brainstorming. It increases efficiency by sharing materials and communal endeavor. Cooperation is expansive and inclusive. It stimulates networking with colleagues with complementary skills, supports a multi-disciplinary approach to problems, and allows more productive use of energy and resources.

One of the institutions of cooperation in science is widely known as the "old boys' network." As an informal system of communication, the old boys' clubs promote their members by providing recommendations and early access to information. One of the advantages of being a student or a postdoc in a prestigious lab is having the opportunity to gain entry into the old boys' network. Unfortunately, women, foreign students, or those who train in labs not considered to be "on the map" have difficulty breaking into this system. In order to be included, members must carry their own weight. If they do not, they are dropped from the network.

The extent of generosity offered by the old boys' network is often limited to giving young investigators a chance—recommending them as reviewers for journals and inviting them to attend select scientific meetings. Members of the network may also provide interested parties with unsolicited preprints of publications. Since the journal literature generally falls a year or two behind the pace of the research, advance knowledge of the latest results can make a dramatic difference to the direction of a research program. The old boys' network can also serve a mentoring function, facilitate introductions to prestigious scientists, and provide a sounding board for discussing experimental results.

Although not everyone obeys it, there is also an unwritten law in science that obligates researchers to provide materials (such as specific antibodies or clones that are not commercially available) that are necessary for someone else to reproduce published experiments. Providing this service can sometimes cost a great deal of money.

While sharing materials and information is fundamental to science, the extent of give-and-take is often calculated, measured out according to what others expect. Researchers may give just enough to stay in the game, holding back the most important data or materials for themselves, and only sharing what they think they can afford or will not miss. They will not give up a part of a project if they think they can get something out of it. Each researcher must find their own niche. In many ways, this kind of sharing is still just another strategy for winning.

The Feminine brings a new element to collaboration. Coming from the perspective of relatedness, researchers approach the world as a network of connections, sharing information to seek and give confirmation and support. The sharing springs from an enthusiasm for discovery and a love for the science. True cooperation thrives on generosity and trust. It leads to a community within which one can nurture ideas and celebrate discoveries. Collaboration based on generosity means giving others useful information that helps them succeed—above and beyond the measured tit for tat.

Like most other researchers these days, Paula Szkody finds it

unusual to publish a paper by herself. Working in a fairly cooperative subfield of astronomy, she generally works on projects with
one to five colleagues and/or graduate students. She loves the exchange of ideas and information, and thrives on interacting with
people all over the world. She finds it personally rewarding and
exciting because they constantly make new discoveries. In her career, Szkody has noticed two distinct levels of cooperation in her
interaction with colleagues:

> Early in my career I worked on a project with a woman and a man on
> different aspects of a new system. After we had each reduced our data,
> the man called me up and asked whether I was going to publish right
> away or wait. I said, "Well, I'm not quite ready yet, so I'll probably
> wait." He said, "Fine." Then a couple weeks later the woman called me
> up and said, "You really should get this out because he is going to
> publish his results." Whereas his approach was just to say, "What are
> you going to do?" and not offer his advice, she said all the papers
> should come out together to receive equal attention. I think a woman
> tends to be more supportive and offer advice to less experienced col-
> leagues. In the man's world it's much more cut and dried—more of a
> focus on the immediate problem than on each individual involved.[37]

One of the touchy issues in collaborations is giving credit: who
to include as an author on a paper, who to acknowledge, who to
credit for materials or ideas. Some investigators feel that acknowledging others dilutes their role, or diminishes their importance.
Each paper reflects a series of decisions regarding how the authors
value the contributions of team members. Recognition also acknowledges the feelings of others. Authors generally are limited to
the principle scientists, with thanks given in a footnote to those
who reviewed the manuscript. Technicians may or may not be
acknowledged; seldom are they authors. Secretaries, glasswash
personnel, animal care workers, and other support staff are rarely
even mentioned. One evening, after watching a movie, I noticed
the number of people recognized in the credits as they rolled
across the screen—everyone from the stars and the director to the
stunt people, grips, gaffers, hair stylists, and caterers received
credit for their work. As a result, the viewer sees what a remark-
ably cooperative effort creating a movie is, and how many people

gave part of their life spirit to the enterprise. Scientists could learn a lesson from this.

Generously giving credit lends an esprit de corps to a group. It gives each member a sense of accomplishment, pride, ownership, and belonging. It nourishes a joy and enthusiasm for the work. Members are energized by a sense of power-with as they brainstorm and work together. At the other end of the continuum, corrosive competition engenders protectiveness, isolation, paranoia, and exclusivity. In some cases, researchers even refrain from revealing all the information required for others to duplicate their work and thus pull ahead of them in the race.

THE PLEASURE OF COLLABORATION

Almost all of the women scientists I interviewed cited collaboration as one of their greatest pleasures. For example, when I asked a cloud physicist what she enjoyed most about science, she said with enthusiasm:

> Best of all is when you're getting somewhere and you're collaborating with someone and you can be excited together. I think that's the most exhilarating. You think about it all the time—you think about it before you go to bed, you think about it at the grocery store. I love collaborating. I also feel that most of the work is infinitely better when collaborating with somebody.

Her collaborations tend to grow organically from mutual interest in a problem. For example, she recalls:

> A collaboration here on campus began when I went to borrow a book from a colleague, and he was puzzling over something. We stood at his blackboard and he said, "This must go like that, it must go like that." And I went back and thought about it and called him back up—I remember distinctly it was a rainy Friday afternoon, and then I spent the whole weekend thinking about it, and by Monday we were collaborating.

A woman engineer finds cooperation essential to her work: "Without cooperation, nothing would ever get done—absolutely nothing. One of my philosophies is, 'Don't try to be a follower or a leader, just be a collaborator.' That's absolutely basic to my very being."

In contrast to the inefficiency of the traditionally compartmentalized and hierarchical interactions between scientists, Davida Teller, a leading international expert in the field of infant vision, illustrates the power of networking laterally. Teller takes pride in fostering cooperation in an international, interdisciplinary group. She said:

> There is an international group of people who all became interested in infant vision at more or less the same time. It's been a remarkably cooperative group, and I like to think I had some influence in making it be that way. Although we've had our rivalries, and they've been painful, we've stayed quite open in terms of communication. I do have a strong sense of a need to cooperate and I think that often is the best approach to progress in science.[38]

Through her cooperative approach, Teller won over the conservative ophthalmic community to the point where they could stop being territorial and look at the possibility that she (as a basic scientist) really might have something to contribute. Teller's visual testing methods are now used in clinics to find out whether an infant's vision is developing normally.

In response to atmospheric scientist Kristina Katsaros's philosophy of generosity and willingness to share, many colleagues bend over backwards to include her in collaborations. These collaborations involve more than exchanging information over the phone or sending materials through the mail. Katsaros and her colleagues often spend weeks at sea together setting instrument buoys and collecting data from them. Many men exclude women from collaborations, either consciously or unconsciously, because they do not feel comfortable working with them. Because Katsaros is sometimes the only woman on the ship, it takes "open-minded people" to want her there. They seek her out because she makes significant contributions to the science. Her collaborations span the globe and extend over decades. In her satellite work, Katsaros collaborates with researchers in France and Scandinavia. Recently, she went to Lisbon for a few weeks to teach for a colleague who helped her with a project in 1979.

Obviously cooperation breaks down when individuals do not carry their own weight. There is also a danger of overspecializing in order to avoid competition, and then holing up in a niche and stagnating. More subtly, cooperation breaks down when the collaborators do not respect each other's talents and efforts. This goes back to Jung's theory of psychological types as discussed in chapter 3. The intuitive type person who stares out the window and receives an intuitive flash, or the feeling type person who listens to the interpersonal problems of the people who work in the lab, often is seen as goofing off. Or sometimes the "grunt work" at the lab bench is valued as less important than the efforts of the one who designs the elegant experiment or analyzes the data. A person walking along the beach reflecting on the data may not be perceived as working as hard as someone scrambling to collect specimens in the field. In fact, analytical insights are no more or less important to a project's success than the efforts of the person who attends to the feelings of the members of the group so that they can continue to productively work together.

Successful collaborators honor the diverse ways their partners contribute. They appreciate others for offering brilliant new ideas and perspectives, providing scarce research materials, sharing expensive equipment or technical expertise, doing the drudge work at the lab bench, keenly observing in the field, analyzing a complex set of data, synthesizing data from diverse sources, or eloquently presenting the results at a meeting.

In the face of today's overriding competition, Sigma Xi declares it has an important role to play. As the honor society for research scientists, Sigma Xi emphasizes the advantages of cooperation through its chapters and clubs that link individuals, departments, and institutions. It encourages its chapters to "act as the catalysts for the recognition that resources are limited, the tasks are enormous, and that what benefits one institution need not necessarily be seen as depriving the other."[39]

Fortunately, many collaborations succeed to everyone's benefit. While competition stalks the area of molecular biology, cooperation is the norm in less populated fields like atmospheric sciences, where there are more than enough problems to go around before researchers step on each other's toes. A cloud physicist tells how they avoided competition and pooled resources at their university:

Just through people talking, we discovered that four or five groups were sending away for the same satellite data, doing a lot of work to process it, and trying to answer similar questions. So we decided to have a little seminar once a week, where one of us gets up and says, "This is what I plan to do, this is how I'm doing it, and this the data I'm going to get." And then other people say, "Oh, well when you get it, I would like it, too," or "Let's pool our resources and hire a computer programmer." Or we've said, "Oh my God, that's what I was going to do. Let's see. How about if you work on this scale and I work on that scale." By pooling our resources we all come out way, way ahead. It's to everybody's advantage, obviously—although sometimes you go to a meeting, and somebody presents a brilliant paper on something that you had thought about but hadn't gotten ready to do. But you hope it also happens in the other direction.

Astronomer Robert Loewenstein from the University of Chicago exuberantly describes a new 3.5-meter telescope (the fourth largest in the continental U.S.) jointly operated by a consortium of five universities. What's special is that this is the first "remote observing" telescope. Users run the entire Apache Point Observatory in New Mexico via telephone modem from Macintosh computers at their home university. Aside from saving considerable time and expense by eliminating the need to travel to a distant observatory, the Apache Point facility promotes collaboration. Although only one person at a time directs the telescope, any number of students and researchers around the country can hook up and "eavesdrop" on the observation. As the observation progresses and the image unfolds on their screens, researchers excitedly talk about what they're seeing, pooling their experience and ideas. In fact, the first operators found that so many messages flew back and forth amongst viewers that they redesigned the format of the viewing screen to expand the dialog space. In addition, they linked up Adler Planetarium so that the public can observe the same displays seen by the astronomers. This way, the public can watch how scientists approach their experiments.

THE DANGERS OF COLLABORATION

Because of the pleasure they derive from collaborating, many women's most significant work has been done with others—

usually men, since women remain a minority in science. This has given rise to the persistent myth that women are incapable of creative innovation. While a woman's natural mode may be cooperative, in collaborative efforts she is often perceived as a hanger-on by those outside the relationship, rather than as a creative and equal partner in the work. Unfortunately, like my friend in the biotechnology company, women may find themselves in a double bind: if they collaborate, they are seen as incapable of independent research; if they work independently, they lose the pleasure and benefits of interacting with others.

This view that women are not capable of innovative work has been expressed openly by some people even in fairly recent years. There's the example I mentioned earlier of a well-known male physicist saying, "If I had been married to Pierre Curie, I would have been Marie Curie."[40] Limnologist Sue Kilham experienced this bias in a very personal way. When she became a widow at a young age, she applied for a position at Drexel University. Although she had been a research professor for eighteen years, the university considered her untried on her own, because her husband had been a famous ecologist. It delayed her tenure until at least two years "on trial."[41]

Under the rules of hierarchy, young assistant professors cannot collaborate with postdocs from big, "high-powered," or prestigious labs—unless the young investigator already has achieved a reputation as a "heavy hitter." In order to get grants and tenure, assistant professors must prove their independence. Collaborations are limited to other researchers who are still "small peanuts." Shortly before cell biologist Aimee Bakken came up for promotion to full professor in the Zoology Department at the University of Washington, she began collaborating with a molecular biologist to augment her expertise in the area of developmental cell biology. Although she had two grants, several postdocs, and several graduate students, this collaboration squelched her promotion:

> Whereas all my male colleagues get good brownie points for having collaborations, I was dinged for it. Just before they were going to take the vote, this guy pipes up, "Well, you know molecular biology is a new field for Aimee, and I really feel like we should wait until she has

198

published some papers without her collaborator's name on it." What can you do? You're collaborating with someone. You can't go publish the data on your own! So I felt like I was really burned.[42]

In some team efforts, many women feel invisible. Marine biologist Rebecca Hoff described a frequent scenario in their lab group meetings. Typically, Hoff and her male colleague discussed their data and interpretations in depth as they worked together. They liked each other and had a good working relationship. In group meetings, each reported their observations. Hoff says with exasperation:

> I would say something and make a point, and there would be absolutely no response. Later he would repeat it and everybody would say, "Oh, wow." He's a big guy and speaks with a loud booming voice—and that's what people tune into. We talked about it and he said the whole male way of communicating is to pretend to be an authority, even if you're not. But I couldn't yell like that. This is just an informal discussion and yet nobody is listening.[43]

Hoff does not have a loud booming voice, but she does speak with a sense of inner authority—far from the soft wispy voice of a little girl. Even though her colleague credited her observations and ideas when he repeated them, she felt frustrated by not being heard the first time. A woman oceanographer also complained, "I want to be heard without discarding common courtesy. I've learned to interrupt and cut people off and I hate it. It's a strain to be fighting constantly to even speak." Similarly, several soft-spoken men said they faced the same conflict.

Scientists have egos and work very hard. They want some credit in the end. Over the years, atmospheric scientist Kristina Katsaros accumulated data on atmospheric radiation from many different latitudes. While these data are like bread and butter to meteorologists and oceanographers, the funding agency only begrudgingly funded it because the science seemed humdrum. One responsible person snidely remarked that Katsaros was providing a kind of service, as though that reduced this work to a menial level. In fact, the same person had asked for her participation to round out the measuring program. Although many scientists make use of the data, Katsaros has to fight for acknowledgment:

People come to borrow data all the time. Sometimes it has happened that they forget to say where it came from—that I don't like, but it happens. Now I've learned to stick up for myself and say, "Please put into the paper where you got it." When they send you a manuscript, you can at least tell them. I'm not too polite anymore. It's amazing how politeness and manners are kind of gone. You hope to be recognized without having to beat your own drum. It's a very touchy point. I've seen people who lost friendships completely—people who should be close to each other, the only two people who understand the same subject.[44]

Caught in the press of competition, several graduate students and postdocs asked me whether cooperation can succeed in the face of competition. Are we being naive to try to cooperate? Won't we just be swallowed up? Unfortunately, students and postdocs are in an unequal power position and can be quickly outdistanced by a big lab machine. It is particularly difficult in "hot" research areas. But even there it may be possible to avoid competition by creating a niche, then communicating with others to negotiate the aspects each will study and the role each will play. Then every person can make a contribution and avoid duplication of effort. Often fears of direct competition are unfounded. If collaborators are chosen with discrimination, and one works and networks with others in an atmosphere of care and trust, cooperation benefits everyone. In selecting collaborators, it is equally important to assess motives and values as well as scientific expertise. While competition is narrowing and isolating, cooperation is expansive. Individuals can start their own experiments with cooperation, beginning one relationship at a time and observing the results. Perhaps by starting a flow of generosity, pockets of cooperation will expand so that we have a dynamic equilibrium between healthy competition and cooperation.

Ultimately, if we derive deep satisfaction from cooperation, and then feel forced to give that up, we lose a part of ourselves. I believe that is a far greater price to pay than being second to publish a result, losing funding, or even losing a job. Some people even refuse to sacrifice their sense of cooperation in order to play the game. Says Davida Teller:

In the cases where cooperation is useful, I usually do very well. In the cases where heads are being lopped off, I'm liable to get my head lopped off, but I still wouldn't have it otherwise. I like myself best when I act like myself, and there's a very strong sense of cooperativeness in me that I would make myself unhappy to put aside. There are times when it might have been an advantage to be a different person than I am, but it wouldn't be worth it.[45]

Jungian analyst Anne de Vore observes that, while men feel unhappy when they lose a contract, women feel hurt and "burned." Personally, I find that competition takes my energy away. It makes me feel distracted, paranoid, and narrowed down. Defending my territory saps my creativity. Although the pressure of competition may make me work longer hours in order to survive or prove myself, I resent it. It erodes my enthusiasm and commitment. Over time, that resentment interferes with the work and ultimately paralyzes me. I can no longer force myself to do it. The consequences to myself and the rest of my life are not worth the price. As long as I played into the competitive dynamic, I compromised myself. Although my first response was to withdraw from the game into isolation, I now know the fulfillment brought by cooperation with people I chose with discrimination.

FRIENDS OF THE UNIVERSE

Based on a sense of interconnectedness and interdependence, the game of science can be restructured from win/lose to win/win by looking to higher goals to which everyone can subscribe. The late prime minister of India, Rajiv Gandhi, spoke of truth and harmony as the larger goals of science:

Generation of knowledge is the test of a civilization. India has long been engaged in understanding the outer and inner worlds and the very process of knowing. It is one of the early fountainheads of science. Indian science is astir again. Through science we are engaged in improving the lives of our people. But our aim is larger—to go to the heart of truth and harmony.[46]

Today, the common goals of repairing environmental damage to the planet and creating sustainable technologies in harmony with nature could replace the arms race as the major emphasis of scientific effort. At the same time, cooperation could ameliorate competition.

201

An example of the expansive power of a cooperative attitude can be found in Buckminster Fuller, who felt his personal success could not be based on the deprivation of others. Fuller, who died in 1983, dedicated his life to bringing humanity and nature into harmony. His impressive contributions to physics, politics, architecture, philosophy, design science, and mathematics were all founded on a sense of cooperation, "because every action has its reaction and resultant, and because no event in Universe can be independent of the rest of Universe. . . ."[47] He wrote that "integrity can no longer tolerate selfishness."[48] Fuller often spoke of spaceship Earth and felt "true happiness could only develop through an awareness that our efforts were always in the direction of progressively-increasing advantage for all humans, without any biases whatsoever."[49] A reviewer of Fuller's twenty-first book, *Critical Path* (1981), called it "a plan designed to bring world opinion and world government to the point of cooperating with nature 'to accomplish what nature is inexorably intent on doing.' Which is to convert all humanity into one harmonious world family, sustainingly, economically successful."[50] After listening to Fuller lecture for eight hours each day for a week, Ezra Pound encapsulated his impression of Fuller in a short poem:

To Buckminster Fuller,
friend of the universe,
bringer of happiness,
liberator.[51]

Avoiding competition and fostering cooperation can help change our attitude from one of using science as a tool of power to control nature, to one of using science as a tool of love to find out how we can better harmonize with nature. At the individual level, scientists can ask, "What unique contributions can I make? How can we work together to solve this problem?" rather than "How can I beat this guy out?" With an active equilibrium between healthy competition and cooperation, not only will women feel more comfortable in science, but men, science, nature, and society will also benefit. We can all become friends of the universe, bringers of happiness, and liberators.

9

INTUITION
Another Way of Knowing

Because it appears mysterious and nonrational, intuition—insight or knowledge gained without evident rational thought—has been associated with the Feminine in our culture. Some scientists deny there is such a thing as intuition, saying it is just a large number of small rational steps that happen faster than we notice. The logician deprecates the fuzziness of intuition and calls it fuzzy thinking. Others, such as Mario Bunge in his philosophical analysis *Intuition and Science*,[1] present intuition as inarticulate visions and hunches that never connect with rationality.

In her book *Awakening Intuition*,[2] psychologist Frances Vaughan describes four levels of intuitive awareness: physical, mental, emotional, and spiritual. While science sometimes embraces physical and mental levels as embodied in scientists such as Albert Einstein and Richard Feynman, it rarely attends to the emotional and spiritual levels. What has been pejoratively labeled "feminine intuition" is intuition on the emotional level. This level of intuition can also help anticipate or resolve interpersonal problems in science. Because girls in our culture are not trained to repress feeling as much as boys, women learn to infer meaning from a range of interpersonal signals often ignored by men. Studies of people's sensitivity to nonverbal communication indicate that women tend to be more attentive to visual cues such as facial expression, body gestures, tone of voice, and the way people look at each other.[3] But intuition is more than keen observation. It is a holistic awareness that includes diffuse sensitivity to both the internal and external worlds, and sometimes even transcends input from the senses.

Some women, such as Ingrith Deyrup-Olsen, defensively balk at the concept of "feminine intuition" because it often has been used to devalue women's thinking. Home-schooled with her six brothers and sisters by her intellectually oriented parents (both of whom had Ph.D.'s), she was never told that girls cannot think or do mathematics. Now in her seventies, Deyrup-Olsen grew up in a generation in which most women were denied scientific education and were constantly subjected to the impact of people saying, "Women can't understand. Women don't know mathematics. You're very sweet, but after all we don't expect you to understand." She said:

> I've always had problems with "feminine intuition." When I was in graduate school, I found myself in this embarrassing situation. I had no problem handling the simple mathematics used in physiology, and most of my male colleagues had difficulty because they hadn't had calculus or differential equations. I actually loved mathematics, so I often found myself in the position where I could explain some mathematical thing very easily to my seniors. On one occasion, when this person had asked me how I'd been able to figure something out, I said in amazement, "How could anyone not understand something as simple as that?" Later, I was berated privately by a senior colleague. He said it would have been better to say, "Well, it's feminine intuition." I thought it was so ridiculous. But the next time it came up, I said it, and everyone just accepted that as if it was the most natural thing in the world. So I have a feeling that people say "women's intuition" because it's more polite than to say that "anyone with half a brain would understand." For me, it isn't a matter of intuition. I go through an intellectual process similar to what the males are doing.[4]

In many such cases, the term "feminine intuition" has been used as a way to explain away women's ability to think. Based on the stereotype of women ruled by instinct and emotion, the reasoning goes: women cannot think, therefore they must have used intuition to arrive at the right answer. In technologically oriented Western cultures, which value rationalism and objectivity, it is generally assumed that intuitive knowledge is more primitive—and therefore less valuable—than so-called objective modes of knowing.

Modern science stresses the importance of empirical data and objective reality (sensation) on the one hand, and systematic, impersonal method (thinking) on the other hand. Consequently, feeling and intuition have been underemphasized because they tend to be perceived as antithetical to the notion of science—as vague, inherently subjective qualities of thinking. Early in this century, the logical empiricists wrote of science as a superior form of knowledge because it is empirically verifiable. Using inductive logic, scientists propose hypotheses and then reject or confirm them by experimental testing. But this construction of science ignores intuition, imagination, or receptivity to new ideas.

Some scientists, however, appreciate that intuition can be a remarkable tool. As one of the Apollo moon scientists said, "I wish I had more of my wife's intuition and less of my own tendency to be reasoning in my approach."[5] In many cases, however, the connection of intuition with the Feminine carries with it a stigma. For example, one immunologist felt his intuition had been the basis for his success because it gives him good ideas, prompts him to follow leads that other people dismiss, and helps him solve ill-structured problems that others tend to shun. But he hesitates to talk about intuition because he equates it with the feminine side of his character. He said, "I can't talk to my colleagues about that because they would think or say that it was too feminine, too nonscientific, or too nonrational." He was afraid others would laugh at him if he admitted he uses intuition to do research. It would not be good for his reputation. While he recognized that his colleagues would not approve or respect it, he thought that the most brilliant biologists he knew were ones who were able to "get in touch with their intuition."

Sara Solla, a theoretical physicist at Bell Labs, believes that finding a door to one's intuition is an integral part of the maturing of a scientist. As a member of an interdisciplinary group working on neural networks—systems that mimic or model brain behavior— Solla plays with analogies as part of her intuitive process. She said,

Most of what I would call flashes of intuition have to do with resonating with something that I have already dealt with or thought about, and

205

analogies—something reminds you of something else and you say, "I thought about that problem in a particular way, maybe the same approach would be fruitful to this problem."[6]

Analogies arise out of individual experience. While two collaborators may work intimately on a given project, the images and analogies integral to their thought processes may be utterly different. Solla finds it amusing that she can be in perfect agreement with her collaborator "but for totally wrong reasons in some sense, having arrived at a conclusion with very different mental processes." The process of intuition is subjective, coming from within.

While many scientists deny using intuition, it is not as deep in the shadow of science as feeling. Mathematicians rely on axioms that are self-evident or "intuitively obvious." Some fields of science, such as those dealing with the theoretical aspects of astronomy or geology, attract more intuitive types than others. In these areas experiments are difficult, if not impossible. For example, 51 percent of the men studying the moon rocks brought back from the Apollo missions believed that rationality and intuition must go hand in hand—that each sustains the other. One said, "A great amount of intuition goes into a good scientist; there is subconscious reasoning. Completely rational people don't make good scientists."[7] Without hunches and intuitions, these men found rationality, logic, and systematic method to be empty and confining. Similarly, they believed that without some form of rationality, wild intuition simply remained unintelligible, aimless, and undisciplined.

INTUITION AS A PSYCHOLOGICAL TYPE

As we discussed in chapter 3, sensation and intuition are modes of perception, where sensation relies on data received through the senses. Intuition refers to the mode of perceiving objects as possibilities. Whereas sensation perceives objects as they are, in isolation, and in detail, intuition perceives objects as they might be and in totality, as a gestalt. The intuitive raises unconscious perception to the level of a differentiated function, by means of an especially sensitive and sharpened perception. Just as we train our eyes to see subtle shades of color, so can we train our intuition to discern and interpret a variety of dimly conscious stimuli. Jung describes intuition as follows:

> It is the function that mediates perceptions in an unconscious way.... In intuition a content presents itself whole and complete, without our being able to explain or discover how this content came into existence. Intuition is a kind of instinctive apprehension, no matter of what contents. Like sensation, it is an irrational function of perception. As with sensation, its contents have the character of being "given," in contrast to the "derived" or "produced" character of thinking and feeling contents. Intuitive knowledge possesses an intrinsic certainty and conviction, which enabled Spinoza (and Bergson) to uphold the *scientia intuitiva* as the highest form of knowledge. Intuition shares this quality with sensation, whose certainty rests on its physical foundation. The certainty of intuition rests equally on a definite state of psychic "alertness" of whose origin the subject is unconscious.[8]

In science, the sensation type is the experimentalist who is guided by the facts and is careful not to extrapolate beyond them. In the extreme, the sensate can become data-bound, forever collecting data rather than risk a generalization that goes beyond the solid facts. With nose to the grindstone, the sensate can spend years verifying trivial hypotheses or doing meaningless research. In contrast, the virtue of intuitive types is that they scan the facts and *do* extrapolate beyond them, envisioning future possibilities. The intuitive function jumps ahead to ask, "What am I going to do with this? Where am I going? What does this mean?" While the sensate perceives the details, the intuitive looks for patterns. Without intuition, researchers competently gather data to fill in the holes or increase precision to the next decimal place, but rarely produce anything new. The intuitive is quick to formulate global, sweeping views of problems and generate a large number of interesting hypotheses. But in the extreme, the hypotheses may be fantasies based on no data at all. This can result in misunderstandings and clashes between the two types, such as occurred with my boss.

As an introverted sensation type, I am sensible and practical, firmly grounded in the real world and in the present. During my scientific training, I developed my thinking function. More recently, I have been developing my feeling function through my psychological work. But intuition remains mostly in my shadow.

According to Jungian theory, the fourth function (the function opposite the primary function in the pairing of thinking/feeling, intuition/sensation), the last to be developed, always remains the weakest. To me, intuition seems magical. I court it indirectly through my thinking and feeling functions. I find myself attracted to intuitive types, fascinated by their imagination, vision, and ability to synthesize and intuit the meaning of a wide variety of experimental and theoretical facts and ideas. I covet the intuitive's talent to perceive inner meaning and significance, to see around corners and beneath the surface. Yet in a work situation, I have the most difficulty dealing with the extraverted intuitive thinker who overwhelms me with an avalanche of ideas and insufficient time and resources to carry them out.

My boss, the extraverted entrepreneur who started the biotechnology company where I worked for five years, used intuition as his primary function and thinking as his auxiliary. A wonderful explainer of science, Bob Nowinski captured the imagination of an audience with his quick wit and ready analogies. Capable of thinking in large, conceptual terms, he rapidly grasped new ideas, saw all the ways they could be applied, and infected others with his enthusiasm. He routinely bounced into our scientific meetings, absorbed the data, proposed experiments and new directions, and then bounded out of the room—leaving us to joke, "Who was that masked man?" I admired his vision and ingenuity. But when it came to the hard work of bringing these ideas into concrete reality, he could not understand why it took so long. He had little respect for the mundane functions of the company and focused his attention on the research groups who presented him with tantalizing new data. He felt dragged down by all it took to bring a product to market. When I presented him with a detailed breakdown of all the tasks and schedules required to accomplish a project, he looked at it in amazement and said, "I could never do this!" His mind moved quickly from one project to the next, seeing all the potential products we could create. He became bored and frustrated by the plodding pace of lab work. Forever at the mercy of a new idea, he leaped ahead of the drudgery onto some new track. By the time the scientists in the company had developed one product, he had already committed us to three more product lines. (This is documented in the book *Gene Dreams* by Robert Teitelman.[9])

A similar conflict occurred between biophysicist Cynthia Haggerty and her supervisor. Haggerty conducted her research in an intuitive way. She simultaneously tried several different approaches to a problem until she sensed which route was going to work best. She then transferred all efforts in the lab to that route. Unfortunately, her supervisor did not understand her intuitive approach. While she had a "gut feeling" about which approach was going to work, she could not articulate her reasons. In an effort to understand and control the work, her supervisor demanded that she write down all her reasoning and justify her choices:

> It absolutely drove him up a wall when I told him my reasons for why this was a good project. He wanted to have all of these linear, logical reasons why we were going to do such and such a project. I didn't want to waste time doing that. I finally quit telling him until I had lots of data. I got into it as far as I possibly could before telling him what I was doing. And I usually had a sense of what would work. That's why we got so much done![10]

Jungian analyst Joseph Wheelwright writes of the conflict between his extraverted intuitive methods of diagnosing patients, and those of a sensation type physician. While on rounds during his medical training, the brilliant diagnostician Lord Horder asked Wheelwright to examine a patient. Although both reached the same conclusion, Horder considered Wheelwright's intuitive methods unacceptable. Wheelwright relates:

> I went over, and there was this man lying in bed with no visible symptoms. I did what seemed to me the most sensible thing to do; I looked out the window for some time. But nothing came to me, so I had a go at the ceiling, and that did it. It all came very clear; there was no doubt in my mind. I turned around with a broad smile and announced in a tone of utmost confidence, "Lord Horder, sir, this man is suffering from pulmonary tuberculosis."
>
> I've never seen a man turn puce so quickly. He fumed at me, "As a matter of fact, he *is* suffering from pulmonary tuberculosis, but that is absolutely irrelevant! How you ever knew it, God only knows. Do you know that there is a diagnostic method that doctors have developed? It

came about because a chap once had some beer in a hogshead and he didn't know the level of the liquid, so he began to thump on the hogshead. We call it percussion. You could have tried that on this chap; you know, we do it on the chest." Then he went on, "I see something sticking out of your pocket; I believe it's called a stethoscope. You could have put that to his chest, and if you were thoughtful, you would warm it first on the palm of your hand. And you would say to him, 'My dear chap, say after me, "*ninety-nine*," ' Did you do any of those things? No! Did you use your voice? You know you might have; we do that quite often. You might have said, 'What brings you here, my good man?' "[11]

Some see intuition as the foe of reason, or as a kind of quackery. Certainly, like any quality carried to the extreme, intuition has its dark side. Nothing is more paralyzing than having an infinite number of possibilities. When not balanced by the other three functions, intuitives can be flighty, impractical, and unrealistic—just as extreme sensates can plod along in a rut, unable to see where all their hard work is taking them. The limited vision of extreme sensates can get so focused on narrowly defined problems and small results that they miss the questions of larger meaning. Because she recognizes the limitations of her type, Marsha Landolt collaborates with an "idea man." Together they form an effective team. She said:

> I have never really viewed myself as a terribly creative scientist. There are certainly things I do very well. I'm a good teacher, a good administrator, a good catalyst. I think one of the reasons my colleague and I work so well together is that he is a wonderful idea man. I'm the person who can get the right mix of people together, get the sponsors on board, track the budgets and the day-to-day operation. In the past when people have said, "Gosh, I hate to see you getting sucked off into administration," and I said, "Look, I do science very much the way I play the piano. When Vladimir Horowitz plays the piano, it's magic. When I play the piano, it's accurate. And there is a far cry between magic and accuracy." I have never been a magic type of researcher. I'm a careful, steady worker. After doing that for twenty years you lose interest.[12]

While intuition leaps ahead into the future, sees wondrous things, and gives us intriguing ideas, it does not take us magically to the goal. The sparkling ideas, tantalizing theories, and vague hunches must be analyzed and prioritized using thinking and feeling. Then to bring the dream to reality, to validate the hunch or theory, it takes hard work in the material world. All four functions—within an individual, a lab, a company, or a discipline—are required to successfully bring a project to completion. Through knowledge of our own psychological type, our talents and limitations, we can more consciously choose collaborators who complement, rather than duplicate, our strengths. Respect and compassion for each type leads to creating the most successful relationships and produces the best science.

COURTING INTUITION

When we want to accomplish more in science, we do more experiments, working faster or longer hours to crank out more data. But this tends to give us more of the same kinds of knowledge. It rarely results in the breakthroughs that come from the "intuition leap." Unlike working harder or organizing better, intuition cannot be pushed or controlled—it must be courted. When we do so, it gives us more choices and thereby more freedom. Unfortunately, the political, administrative, and economic pressures of modern science make it difficult to enter the relaxed, playful state of mind that fosters the intuitive process. Theoretical physicist Eberhard Riedel covets the time when he can afford to let his mind float in a state of timelessness, without being jolted back to reality by interruptions:

> I resent being forced into a mold that keeps me busy busy busy all the time without time for "play." I feel I'm most productive in my research when I have stretches of uninterrupted time. I get frustrated when I get rolling and then have to go to a meeting. Then the meeting looks like the enemy. It's too much for me. I dread the politics of the research community competing for funds. They get into stabbing the knife when there is less funding.[13]

Like all characteristics of the Masculine or Feminine, intuition is a human potential that anyone can develop. While feeling is not considered very manly in our culture, intuition is somewhat more acceptable. As with the Apollo moon scientists, many researchers accept intuition as complementary to their thinking function. Although thinking and sensation may be the primary and first auxiliary functions of most scientists, the theoretical side of science also draws the intuitive thinking type.

Perhaps intuition has been identified with the Feminine because it requires a receptive state of mind. Those who have studied the intuitive process have observed four phases to the intuitive process: (1) Preparation or the "input" mode, in which a person directs a question to the unconscious and provides it with information. This is an intense period of conscious thinking, reading, and research. (2) Incubation or the "processing" mode, where the accumulated information simmers in the unconscious. All input has ceased and this is a time of relaxation, daydreaming, meditation, or sleep. (3) Illumination or the "output" mode, when a mysterious process produces the solution to the problem in a flash, seemingly from nowhere. (4) Verification mode, the phase in which fantasies are discerned from inspirations, and delusions are distinguished from insights.[14]

Paul Hamilton, a systems analyst for Boeing Computer Services, relies on his intuition for solving problems. While he knows he cannot force his intuition, he has learned how to promote its occurrence. He describes his method as skittering on the surface of his mind. Like a low level of meditation, he keeps his conscious mind quiescent and focuses on a question. Although aware of random thoughts, like wondering if his tie is on straight, he waits attentively for an answer to arise. While he cannot control it, he perceives it coming, like a sneeze gathering. Analytical thinking stops it. Once he has the flash, then he goes back and builds a chain of logic to explain his "reasoning" to colleagues. He finds that the demands of external reality stimulate the activity of creative fantasy. After the initial burst of illumination comes a dynamic interplay between intuition and analysis. Because he has nurtured his intuition and constantly practices using it, he trusts it and depends on it.

A woman engineer attributes her success to her intuition. She observes:

> There are two types of engineers: there's the kind that you usually think of, the engineer who is very good at details—the sensation type who

sits down, concentrates, and goes step by step, one job at a time. I'm entirely different. I really am not a good detail person. I can concentrate as much as I have to, but what I'm really good at is handling a dozen different projects at once without being flustered. What I do is go over it, see what it's all about, put it on the back burner of my mind to percolate, and then very often the one answer (or the way to get the answer) just pops out—in the middle of the night in a dream, or whatever. I do that all the time and it's a fairly rare quality that comes from being intuitive. I'm sure that's why I got promoted instead of anybody else in the group. It's not what you generally think of as an engineering ability, but it's actually very important. This is more the creative side of engineering. A lot of people think engineers are just mental drudges, trudging through page after page of calculations. That's the one type of engineer. The other type is the creative type who is more like the scientist—the insightful type.

The flash of intuition tends to emerge during times of relaxation—in the shower, in a dream, while taking a walk, while staring out the window. It eludes the grasp of the tight, achievement-oriented mind-set. Trauma, worry, and stress also inhibit the process. The climate of tension and anxiety created by the oh-so-serious business of science suffocates the relaxed, playful, exploring intuitive mind. Rationalists will often have good ideas, but then think they are nonsense and press on with their rational planning. In today's competitive climate of science, immunologist Julie Deans feels constant pressure to produce. She misses the more reflective, meditative days as a graduate student. "There's not enough time to be creative, what with conferences, writing papers, and constant work at the bench. I never seem to have time to stop to wonder. It's harder to drop back and punt as I get busier and busier. I need more time to think to allow ideas to bubble up."[15]

As the adage "seeing is believing" tells us, we are more likely to trust the information we receive through our senses than from our intuition. In reality, however, both modes of perception can trick us. Neither sensation nor intuition are rational or evaluative functions. Sensation gives us information about the world. Intuition reveals possibilities and provides insight into the nature of things. But neither substitute for reasonable evaluation or moral consideration of the information we receive. Just as thinking and feeling functions need feedback from the perceptive functions, so do sensation and intuition need to be evaluated by the rational functions of thinking and feeling. True perception must be discerned from self-deception and wishful thinking. Jung stressed that we must never passively accept the revelations of our intuition as absolute truth, but rather we must interact with them, raise questions, and present objections in a spirited dialogue.

INTUITION EMBODIED

Albert Einstein, so totally wrapped up in his imagination that he walked around unconscious of the daily world of sensation, exemplifies the intuitive scientist. He routinely lost keys and gloves; he often did not think to wear a coat or comb his hair. One day, while on a walk around the streets of his Princeton home, he forgot where he lived. The great discoveries of Einstein's earlier years were all based on direct physical intuition. At age sixteen he had the idea that was to bring about a revolution in physics. He said, "It [the optics of motion] occurred to me by intuition. And music is the driving force behind this intuition. My parents had me study the violin from the time I was six. My new discovery is the result of musical perception."[16]

Einstein separated himself from the logical positivists by repeatedly expressing his reliance on intuition: "There is no logical way to the discovery of these elementary laws. There is only the way of intuition."[17] "[O]nly intuition, resting on sympathetic understanding, can lead to [these laws]. . . . The longing to behold [cosmic] harmony is the source of the inexhaustible patience and perseverance. . . . The state of mind that enables a man to do work of this kind is akin to that of the religious worshiper or the lover; the daily effort comes from no deliberate intention or program, but straight from the heart."[18] Rather than manipulate physical objects, he "saw" mathematics. He could playfully reproduce and combine the objects of his imagination as easily as external visual objects. He wrote, "The objects with which geometry deals seemed to be of no different type than the objects of sensory perception 'which can be seen and touched.' "[19]

Physicist Richard Feynman also embodied the playful freedom

214

of the intuitive scientist. He argued that Einstein had failed because "he stopped thinking in concrete physical images and became a manipulator of equations." While Einstein's early discoveries were all based on intuition, his later unified field theories were, according to Feynman, only sets of equations without physical meaning.

A brilliant theoretician, Feynman received the Nobel Prize for Physics in 1965 for work in redefining the basic principles of quantum electrodynamics. His friend, physicist Freeman Dyson, characterized his intuitive approach by saying that a comment about Newton applied equally well to Feynman. "His peculiar gift was the power of holding continuously in his mind a purely mental problem until he had seen straight through it. I fancy his preeminence is due to his muscles of intuition being the strongest and most enduring with which a man has ever been gifted."[20]

Dyson served as Feynman's sounding board as he developed his unifying vision of the interactions between radiation, electrons, and positrons. In a profoundly original approach, Feynman constructed simple diagrams to graphically represent interactions of particles. These ingenious diagrams provided an easily visualized analogue for complicated mathematical expressions. Initially, no one but Feynman could use his theories because he was "always invoking his intuition to make up the rules of the game as he went along."[21] Over the past thirty years, however, these diagrams— now called Feynman diagrams—have been absorbed into the fabric of physics. Dyson describes Feynman's intuitive style:

> The reason Dick's physics was so hard for ordinary physicists to grasp was that he did not use equations.... Dick just wrote down the solutions out of his head without ever writing down the equations. He had a physical picture of the way things happen, and the picture gave him the solutions directly with a minimum of calculation.... The calculation ... using orthodox theory, took me several months of work and several hundred sheets of paper. Dick could get the same answer, calculating on a blackboard, in half an hour.[22]

> He has developed a private version of the quantum theory, which ... may be more helpful than the orthodox version for some problems; in general he is always sizzling with new ideas, most of which are more

spectacular than helpful, and hardly any of which get very far before some newer inspiration eclipses it.[23]

In 1986, Feynman became known to the general public as a member of the presidential commission investigating the explosion of the space shuttle *Challenger*. While others members of the commission read documents, attended meetings and hearings, Feynman ran around the country checking all kinds of unusual things. Through informal talks with technical people, he rapidly achieved an understanding of the shuttle's systems. Rather than systematically examining all the departments of NASA, he talked to assembly workers and ate lunch with the engineers. He listened to the concerns of the guys at the bottom as well as the "big cheeses." Through his unorthodox probing, he found that the technical people knew of dangers and problems that the managers ignored or covered up. When Feynman became bored by the formal "briefings," he made up games for himself, imagining what the commission might discover if a different system had failed. He wondered if they would find the same slipping safety criteria and lack of communication. Instead of simply filing a technical report full of jargon, Feynman also performed a dramatically simple experiment at a public meeting. He placed a piece of the shuttle's crucial O-ring seal in a glass of ice water and squeezed it with a small clamp. When he removed the clamp, the seal failed to spring back into shape, demonstrating that the rings were not resilient enough to maintain their shape after being subjected to the cold conditions at the launch site.[24]

Feynman's intuition imbued his life with an enviable freedom and sense of adventure. He played the bongo drums and began taking art lessons at age forty-four. He enjoyed traveling to places in the middle of nowhere, places he had never heard of but had an intriguing name, places no one else thought to visit. His intuition gave him a flexibility and broadened the available choices. For example, when the owner of a Japanese-style inn in Iseokitsu was reluctant to accommodate foreigners because the inn did not have a Western-style toilet, Feynman was undaunted. He said to the person making his arrangements, "Tell them that the last time my wife and I went on a trip, we carried a small shovel and toilet paper, and dug holes for ourselves in the dirt. Ask him, 'Shall we bring our shovel?' " The innkeeper acquiesced saying, "It's okay. You can come for one night. You don't need to bring your shovel." Feynman's intuition saw the possibilities in places others said had "nothing." In Iseokitsu he found beautiful gardens, an emerald green tree frog, a shrine, and generous people. He played with the innkeepers' young children and the family responded to his warmth and sense of fun. He and his wife cancelled reservations at a resort hotel and stayed a second night at the inn.[25]

CREATIVITY

The Void or Chaos is the starting point for nearly all creation myths. In the Hindu story of creation, the sun and the moon and the animals emerged from churning the milk ocean. In many cultures such waters are a symbol of chaos and of the unconscious—that reality where all things rest in potentiality. Likewise, Prigogine's idea of a creative void, of chaos as the generative source of the universe, is similar to Jung's (and the alchemists') view of the unconscious. While logic and analysis emphasize predictability, the nonrational process of intuition is unpredictable. After laboriously filling our mind with information and struggling to find the answer to a perplexing problem, a flash of intuition sometimes provides the solution. Just as patterns emerge from the nonlinear dynamics of chaos theory, intuition appears to arise mysteriously from the unconscious to provide us with new information. In nonlinear systems, logic cannot serve to predict. However, with intuitive awareness of the whole pattern, we can make decisions based on a sense of the whole, rather than merely extrapolate from the past. Intuition gives us flexibility and the ability to respond spontaneously to change. Just as complex systems move toward increasing levels of complexity, and systems engage in spontaneous self-organization, so does intuition spontaneously arise from the dark waters of the unconscious and provide a source of renewal.

While artists speak freely of the creative process, scientists are more likely to describe their work as "revealing what is." Biologist Peter Medawar observes, "Scientists are usually too proud or too shy to speak about creativity and 'creative imagination'; they feel it to be incompatible with their conception of themselves as 'men of facts' and rigorous inductive judgments.''[26]

Although some scientists know that the truly great breakthroughs come from intuition, they rarely study how intuition works. As far as most scientists are concerned, it is a mechanical process of the brain that, if you put in enough information and stir it around enough, a new combination will come out. While it is true that insights emerge from reorganizing information in different ways, that does not constitute creativity. The creative leap goes beyond the existing information and adds something new to it, pulling it seemingly out of nowhere. And we still do not know how that functions.

Sometimes that flash of illumination comes literally from dreams. In their book *Higher Creativity: Liberating the Unconscious for Breakthrough Insights,* Willis Harmon and Howard Rheingold draw together numerous cases where artists and scientists attribute their creative breakthroughs to images that arose in their dreams.

The most-cited example of dream-inspired science comes from the nineteenth-century chemist Friedrich August Kekulé von Stradonitz. While dozing in his chair by the fire, Kekulé dreamed of a snake biting its tail. He awoke "as if by the flash of lightning" understanding the ring structure of benzene, a problem that long had eluded chemists. This vision provided the clue to a discovery that has been called "the most brilliant piece of prediction to be found in the whole range of organic chemistry."[27] Other visions came to him during reveries while riding aboard a streetcar. In a lecture toward the end of his career, he exhorted his audience—a society of chemists—to "learn to dream."[28]

But Kekulé's experience was not an isolated example. Physicist Niels Bohr dreamed of a planetary system as a model for atoms, leading to the "Bohr model" of atomic structure and a Nobel prize.[29] Early this century an uneducated teenage boy from a poor family in India was given an old mathematics textbook. Srinivasa Ramanujan recounts that soon after he read this book, the goddess Namagiri presented him with formulas in his dreams. From these dreams, he constructed a massive body of mathematical knowledge and managed to secure a scholarship at Cambridge, despite his ignorance of nonmathematical subjects. He arrived in England to find his work was often ahead of contemporary mathematical knowledge.[30]

After an intense period of work and concentration, Dmitri Mendeleev conceived in a dream the periodic table that orders all the chemical elements:

> In 1869, D.I. Mendeleev went to bed exhausted after struggling to conceptualize a way to categorize the elements based upon their atomic weights. He later reported, "I saw in a dream a table where all the elements fell into place as required. Awakening, I immediately wrote it down on a piece of paper. Only in one place did a correction later seem necessary."[31]

Elias Howe worked intensely for several years to invent the lockstitch sewing machine. His earlier models failed, and he did not succeed until after he awoke one night from a nightmare. His dream has been recounted in a book on inventions:

> Howe made the needles of his early failure with a hole in the middle of the shank. His brain was busy with the invention day and night and even when he slept. One night he dreamed . . . that he was captured by a tribe of savages who took him prisoner before their king.
>
> "Elias Howe," roared the monarch, "I command you on pain of death to finish this machine at once."
>
> Cold sweat poured down his brow, his hands shook with fear, his knees quaked. Try as he would, the inventor could not get the missing figure in the problem over which he had worked so long. All this was so real to him that he cried aloud. In the vision he saw himself surrounded by dark-skinned and painted warriors, who formed a hollow square about him and led him to a place of execution. Suddenly he noticed that near the heads of the spears which his guards carried, there were eye-shaped holes! He had solved the secret! What he needed was a needle with an eye near the point! He awoke from his dream, sprang out of bed, and at once made a whittled model of the eye-pointed needle, with which he brought his experiments to a successful close.[32]

Otto Loewi received the 1936 Nobel Prize in Physiology and Medicine for his discovery that nerve transmission is both a chemical and electrical event. During a conversation early in his career he had a hunch that nervous impulses might be chemical as well as electrical, but he could not think of an experiment to validate his hypothesis. Seventeen years later, an experimental procedure to test his idea emerged in the form of a dream. Loewi wrote:

> The night before Easter Sunday of that year [1920] I awoke, turned on the light, and jotted down a few notes on a tiny slip of thin paper. Then I fell asleep again. It occurred to me at six o'clock in the morning that during the night I had written down something important, but I was unable to decipher the scrawl. The next night, at three o'clock, the idea returned. It was the design of an experiment to determine whether or not the hypothesis of chemical transmission that I had uttered seventeen years ago was correct. I got up immediately, went to the laboratory, and performed a simple experiment on a frog's heart according to the nocturnal design. . . . Its results became the foundation of the theory of chemical transmission of the nervous impulse.[33]

Few modern scientists publically reveal the sources of their inspirations. An exception is Jonas Salk, who developed the Salk vaccine for polio. Thirty years ago he founded the Salk Institute for Biological Studies to "create a crucible for creativity."[34] Salk keeps a pen and paper by his bed and often awakes with a start to begin what he calls his night writing. In a trancelike state he fills page after page with ideas that seem to emerge from another realm. Now seventy-six years old, he has collected over 12,000 of these nocturnal revelations.[35]

INTUITION AS A FRONTIER OF SCIENCE

Because intuition is unpredictable and individual, coming as a whole in a flash, it cannot be broken down into its component parts for us to study. As yet we do not have the vocabulary or ability to discern the workings of intuition from other types of psychic phenomena such as dreams, telepathy, precognition, and remote viewing. The study of such phenomena calls for a different approach to research. New methods must be found to handle reports of subjective experiences, to cope with the difficulty of replicating psychic phenomena, and to deal with the uniqueness of individual experience. Rather than discount the functioning of the psyche, Jung maintained that:

It is an almost absurd prejudice to suppose that existence can only be physical. As a matter of fact, the only form of existence of which we have immediate knowledge is psychic. We might well say, on the contrary, that physical existence is a mere inference, since we know matter only in so far as we perceive psychic images mediated by the senses.[36]

Research into the fields of human consciousness, psychic functioning, or parapsychology provides a way to bring intuition to consciousness. Unfortunately, research into these areas has been confused by the presence of frauds. Quack researchers make extravagant claims, spending more time in interviews with the press than on careful experiments. Sensationalistic journalism and movies portray psychic abilities as crazy, aberrant, or nonhuman. Con artists exploit popular misconceptions and public fascination with psi, using their real or faked psychic abilities as a source of profit, power, and control. Many cults with charismatic leaders use claims of supernatural abilities to deceive and exploit their followers. They lure people by promising to teach them psychic abilities that will give them influence over other people. Because of such cults and frauds, the public tends to associate all psychic phenomena with social deviants, scams, and charlatans. Together, the frauds and the media frighten away many people who are interested in learning about our human psychic potential.

In addition, many of the critics of psychic research are motivated by their own private fear of the unknown. Most are extreme materialists who believe that if we do not understand something, it cannot exist. Such rational scientists eschew as superstition anything they cannot measure with their five senses. Acceptance of the validity of some psychic phenomena requires us to redefine and expand our current understanding of physics and psychology. Not everyone is ready for that. Such fears have fueled attempts to drive the Parapsychological Association out of the American Association for the Advancement of Science. (Formed in 1957, the Parapsychological Association was admitted to the AAAS in 1969.) Critics have even fraudulently conspired to discredit research results that did not fit their worldview. Astrophysicist Dennis Rawlins, one of the founders of the Committee for the Scientific

Investigation of Claims of the Paranormal (CSICOP), exposed CSICOP's fraudulent statistical manipulations of unexpectedly positive research findings. While still skeptical of "occult beliefs," he wrote that he has changed his mind about the integrity of the debunkers, and realizes they will go to extreme measures to disparage the credibility of such research "for the public's own good."[37]

No one—least of all a scientist—wants to appear gullible or foolish. Consequently, the association of psychic phenomena with frauds and sensationalism has kept people from taking psychic experiences and abilities seriously, and has inhibited them from learning about the actual scientific work in this field. Just as congenitally blind people who regain their sight after a surgical operation must learn how to interpret visual perceptions, so must we learn how to interpret the subtle and shifting perceptions of intuition or other psychic phenomena.

I personally have known several people with psychic abilities, including a computer scientist in the field of artificial intelligence and the international marketing director of a timber company. Neither are members of cults, nor do they use their abilities for financial gain. They are intelligent, articulate, stable people who lead successful private lives in mainstream corporate America. Their psychic abilities form an integrated part of their awareness. Knowing these people has forced me to question the completeness of our current understanding of time and space.

Over more than a decade, the U.S. government has quietly supported both military and nonmilitary research into psychic functioning. A multimillion-dollar program at SRI International (formerly Stanford Research Institute, now a huge high-tech research institute) explored ways of increasing the accuracy and reliability of a type of perception they called "remote viewing" (the ability to describe locations, events, or objects that cannot be perceived by the usual senses because of distance).

At a remote-viewing session, the "viewer" sits in a comfortable room. A second person, the "beacon," uses an electronic random number generator to select an envelope containing the location of one of sixty possible target sites. The beacon does not open the envelope until in the car. The beacon drives to the site and, at a prearranged time, gazes intently at the site for fifteen minutes.

Meanwhile, the viewer relates impressions and sketches images for the "interviewer," who is also unaware of the target site. At the end of a series of remote-viewing trials, an independent evaluator visits the sites and chooses the description best matching that site. In trials carried out at SRI, approximately two-thirds of the remotely viewed descriptions were successfully matched by the evaluators. The odds of this happening are about one hundred to one against chance.[38] In laboratories across the world, twenty-three series of investigations have reported statistically significant data supporting remote viewing.[39]

Successful trials included long-distance viewing from Detroit to Rome, Italy, experiments conducted in a variety of electrically shielded rooms, and trials that blocked extremely low-frequency (ELF) radio waves by submerging the viewer in a submarine. None of these conditions reduced psychic functioning.[40] In another variation of the experiment, the viewer sketched the site *before* the envelope was randomly chosen. This precognitive viewing also gave statistically positive results.[41] The SRI researchers concluded that remote-viewing is a trainable skill, latent in each of us. They find the strange occurrences of psychic functioning to be rather commonplace, although as yet poorly understood.

This research taken seriously implies the availability of a channel of perception or communication in which time and distance seem to pose no barriers. Like the Aspect experiments discussed in chapter 5, acceptance of intuition and other forms of psychic functioning threatens our concept of locality, a premise of the special theory of relativity that information or forces (such as gravity) can only travel between bodies at velocities less than the velocity of light. The precognitive studies also lead us to question our beliefs about causality, the relation between a cause and its effect—that for one event to cause another, it must occur before the event it causes. Such research also makes one wonder how consciousness is related to other aspects of the world.

Acceptance of intuition gives us greater access to information, augments the limited perspective of our five familiar senses, and prompts us to transcend our linear view of time and space. Intuition can help bridge the boundaries that seem to separate us from others and from nature. Frequently, people who develop their psychic potential feel a strong sense of unity with the natural world. They feel a personal sense of interconnectedness that physics has revealed at the quantum level. But their sense of harmony with the world is not theoretical; it is experiential. They intuitively know that there is something more, something greater for us to reach toward, something deep within ourselves, deep within the fabric of the universe.

While the combination of thinking and sensation produces materialism, the alchemical union of intuition and sensation gives a person the ability to obtain intuitive insights simultaneously with his or her perception in the outer world. The senses become merged with intuition without having to go through the intervening process of thinking or feeling. The act of sensation by way of the senses becomes transfigured and uplifted by way of an intuition that attaches itself to sensation. Then physical phenomena become the openings to intuition. The intuition enlivens and ensouls the physical world, so that every sight and sound brings a titanic explosion of meaning.

When we deny our intuition, we deny our sense of relatedness— to others, to nature, and to our inner selves. While scientific knowledge often stays abstract and theoretical, through intuition we can experience the wholeness of nature and learn to live with it in harmony.

——10——

RELATEDNESS
A Vision of Wholeness

One of the most highly developed skills in contemporary Western civilization is dissection—splitting things up into their smallest possible components. We are good at it—so good that we tend to forget to put the pieces back together again. The Feminine, on the other hand, tends to view each part in context, as part of a larger picture. Our intuition can give us a vision of wholeness.

As the "principle of relatedness," all the qualities of the Feminine arise from a sense of interconnectedness. Feeling, nurturing, receptivity, cooperation, intuition—all are based on interdependence, a keen awareness of relationship to the other and to the whole. In contrast, science has pursued the masculine path of logic and analysis based on separating and compartmentalizing the parts. This path has great power and has produced the marvels of modern technology, but it also has led to problems such as environmental pollution. Used in conjunction with the Feminine, the researcher can focus on the individual parts while simultaneously considering their relationship to the environment. The feminine principle of relatedness can give science a more holistic perspective.

In the process of writing this book, I reflected on my own shift away from the mechanistic world view. My training in science followed the traditional reductionistic approach. Looking for the fundamental basis of life, I studied biochemistry in graduate school. Then, to understand interactions of biochemicals at the molecular level, I studied quantum mechanics and molecular orbital theory, and minored in physical chemistry. In a sense, my personal history recapitulated the course of modern science, looking at ever smaller bits to explain the magic of our world, and hoping that, by taking apart cells and studying them, I could understand Life.

Curiously, in the midst of my plunge into the depths of matter, with its myriad of subatomic particles, I chose a research project that did not lend itself to the reductionistic approach alone. At an unconscious level, my feminine side was resisting the purely analytical path. My feelings led the way. I took a personal interest in the tragic stories of people who inherited a tendency to react to anesthesia by a dramatic (often fatal) rise in body temperature. Theories of the mechanism of anesthesia proposed that anesthetics slowed down nerve transmission. None of the theories, however, accounted for the diverse side effects of anesthetics, such as this increase in body temperature. I demonstrated that anesthetics slow down the rate and strength of beating of individual heart cells in tissue culture—cells free of nervous connection. I had taken the reductionistic approach by looking at individual cells, yet my next step was to relate this finding to the whole body by asking, what did the beating mechanism of heart cells have in common with the nervous transmission? I found that calcium-binding proteins, regulatory proteins that control a number of functions in the body, modulated both mechanisms. My research continued as a dance between the molecular, cellular, and whole body levels.

Jungian psychology attracted me because of its emphasis on wholeness. The goal of the psychological work is not to simply adapt to societal norms, nor to attain a static state of perfection. Rather it is to develop all our potential, to live fully and deeply, to bring into relationship all the opposites of our psyche, to affirm life in all its richness and ambiguity. With the goal of achieving wholeness of the psyche, the process of Jungian analysis consists of examining and accepting parts of ourselves we have denied and restoring communication and flow between parts of our lives we have split off into airtight compartments. Unlike ossified perfectionism, wholeness is a dynamic interaction of the opposites like the pulsing between yin and yang in Taoism.

In contrast to modern science, which takes things apart and analyzes them ad infinitum, Hildegard of Bingen embodies the feminine approach for me. The most prominent scientist of medieval times, Hildegard was in love with nature. She saw interconnectedness and interdependence as the very stuff of the universe. She wrote, "Every thing that is in the heavens, on the earth, and under the earth, is penetrated with connectedness, penetrated with relatedness."[1] Hildegard did not compartmentalize her life. In her work she linked science, spirituality, and art. She approached science as a passionate knower.

Eight hundred years later, Edgar Mitchell describes a similar feeling of connectedness in the universe. As one of the Apollo astronauts, he stood on the moon and looked back at our magnificent little blue and white planet, thrilled by seeing this dwelling place for life, for awareness, and for being:

> On the way home I turned my attention to looking at Earth and the cosmos. Unexpectedly I experienced an exhilarating sense that I and the universe were one—that it is but an extension of myself, that each of us is an integral part of the same existence. That was a very heady experience for me—exciting, joyous, and perplexing. For the first time, I recognized that I was looking at an organism. At about the same time, Jim Lovelock startled the scientific community with his Gaia hypothesis. His hypothesis described how I perceived Earth from space—and how I see the universe as a whole—functioning as an organism, not inanimate, dry, inexpressive matter, as we in science have thought.[2]

Founder of the Institute of Noetic Sciences, Mitchell cites objectivity as one of the flaws of science. The other flaw he sees is reductionism, reducing nature to smaller and smaller parts in order to analyze them:

> I see two fundamental flaws in the scientific method. One is objectivity.... The other fallacious assumption of science is that, if you reduce everything to its ultimate minuscule particle, you can then put it back together and understand the whole thing. Of course we know that isn't true. We live in a universe that is not linear—a universe that adds elements of complexity and beauty the more it comes together into a holistic organism. At every level of growth and organizing, new attributes appear that were not at all predictable or understandable from

227

the elements themselves. You have to look at the whole organism to understand it—not simply at its parts.[3]

Many scientists, however, still cling to Sir Isaac Newton's model of the perpetual, mechanistic, clocklike universe. Even Einstein, on his deathbed in the 1950s, was loathe to let go of the static model of the perpetual universe. The ultimate goal of physics remains the identification of the fundamental particles or fields, and understanding how they interact. These physicists hope to formulate a Theory of Everything, to explain the universe in a simple formula we can wear on a T-shirt. Leonard Susskind, a theoretical physicist at Stanford and SLAC, expresses this nuts-and-bolts view of the universe:

> I think of myself as a mechanic rather than a philosopher. And what we have here is a very large car. We don't know how it operates, so we tinker with it. We push electrons this way, we push them that way, we use accelerators. Eventually we hope to figure out the rules by which this car operates.[4]

There is no denying the usefulness of isolating and dissecting bits of nature. It has led to spectacular discoveries and advances in our knowledge about the world. We have discovered viruses and bacteria that cause disease, created new materials such as plastics, and put men on the moon. Machines ease much backbreaking labor and amplify a person's power; sophisticated construction equipment is used to build dams, skyscrapers, and airplanes. But the mechanistic approach alone is no longer sufficient. Many problems in science have been singularly unresponsive to this approach: curing cancer, predicting earthquakes, long-term weather forecasts, the function of the central nervous system, aspects of animal behavior, developmental biology, and the evolution of mind and consciousness.

Perhaps the most important contribution the Feminine brings to science is a vision of wholeness. This preference for wholeness derives from the fundamental principle of relatedness characteristic of the Feminine. Relatedness means looking at the relationships between things, viewing things in context, seeing the connections that link everything together, stepping back to see the big picture—and even weaving together work and personal life. In doing so, we find the whole gives meaning to the parts. The whole takes on functions that are not suggested by the parts. Theoretical physicist Paul Davies believes that reductionism denies the reality of higher levels of organization such as a biological organism: "As matter and energy reach higher, more complex, states so new qualities emerge that can never be embraced by a lower-level description. Often cited are life and consciousness, which are simply meaningless at the level of, say, atoms."[5] The feminine vision of wholeness sees that each new level in the development of matter brings its own laws that cannot be reduced to those at lower levels. Organizing principles such as cooperation, or the collective properties of complex systems, are overlooked by reductionist methods, since they cannot be derived from underlying existing physical laws. A holistic approach to science can provide a valuable complementary approach to the traditional reductionistic approach.

Like all aspects of the Feminine, the vision of wholeness has always been present in science, but it still plays only a minor role. For example, the most heavily funded projects in science today are justified by the reductionistic assumption that, if the fundamental pieces are understood, then all the rest of nature will be explained. These projects include identifying the fundamental particles with particle accelerators, and sequencing the human genome.

BALANCING REDUCTIONISTIC AND HOLISTIC APPROACHES

While the reductionistic approach is powerful, it could serve us better if married to the holistic perspective. It is the one-sidedness that is the problem. The other extreme—that of science without analysis—would be no better. Then we would be overwhelmed with an impenetrable web of connections. We need the analytical process to help us disentangle the threads, while always keeping one eye on the overall pattern to see what our meddling is doing. As in any endeavor, there is always the question of balance and perspective. Which viewpoint rules our thinking? Which do we constantly seek? Do we see the world as fundamentally composed of basic building blocks—or as a seamless whole? In viewing the interplay between the reductionistic and holistic approaches, physicist David Bohm makes an analogy to music:

> When you're playing music it is very important which theme is given the dominant role and which the secondary role. If we reversed their roles, it would be an entirely different piece. What has happened is that the dominant role has been given to this theme, namely the partiality of the whole and the parts. I'm proposing that we keep this as a secondary role and assign the other a major role.[6]

Sylvia Pollack is a full professor at the medical school of the University of Washington in the Department of Biological Structure. Hers is an unusual department in that half of the full professors are women. A cellular immunologist, Pollack is interested in studying the development of lymphocytes, the white blood cells of the immune system. She wants to know what controls their development, and how one type of cell matures into another type of cell with a different shape or function. She constantly tries to balance the holistic and reductionistic approaches:

> Because what I work with is so complex, I wrestle all the time between being reductionist, which I see as being quite masculine, and being holistic or integrative, which I see as more feminine. I know I'm not going to get anywhere if I'm not reductionist. I've got to chop it into little pieces and look at it. At least I always try remember we're looking at the pieces so we'll understand the whole system. Since I'm interested in how blood cells form, anything that happens in that process is fair game.[7]

Although Pollack finds it frustrating to work with whole systems, she ultimately finds it to be more exciting because of all the possibilities. Studying whole systems challenges her to be more clever and more integrated in her thinking:

> You control as many things as you can, but the animal is always doing things you can't control. If you deal with anything above the molecular level you deal with that. I had a graduate student who just couldn't work in our lab. He really needed to be an honest-to-goodness molecular biologist dealing with a very discrete set of questions and things that he could work with—reagent grade nature. What we were working with was far too murky for him.[8]

Ingrith Deyrup-Olsen takes great delight from her work in studying slugs. Although her methodological approach is reductionistic, her overruling perspective is to think of the position of the animal in its whole environment:

> I have always been interested in things such as the fascinating area of how fetal development affects physiology and the changes that occur at birth. I like to think of the position of the animal in its whole environment. I think my work has tended to end up being very reductionist, because of the kinds of methods that I have and my background. But my enjoyment of the field, and my participation as a teacher, has been very much affected by the holistic idea that I feel I have the right to think about problems in a broader context—even though I can't do anything about it with my techniques in my particular system.[9]

Now let us explore how this principle of relatedness is reflected in newly emerging sciences such as ecology, holistic medicine, chaos science, and quantum theory.

ECOLOGY, THE SCIENCE OF RELATEDNESS

Viewing the world as a machine leads to feelings of separateness, and results in exploitation of nature. This reductionistic approach contrasts with the unifying view provided by ecology, the science of the interrelationship of organisms and their environments. As a multidisciplinary science, ecology embodies the holistic outlook characteristic of the feminine principle. It includes plant and animal ecology, population dynamics, behavior, evolution, taxonomy, physiology, genetics, meteorology, pedology, geology, sociology, anthropology, physics, chemistry, mathematics, and electronics. Applied areas of ecology include wildlife and range management, agricultural production, and addressing the problems of environmental pollution. Long relegated to second-class status by many in the world of science, ecology has now emerged as one of the most important aspects of biology.

Biologist Rachel Carson challenged the notion that science belongs in "a separate compartment of its own, apart from everyday life." To Carson, the aim of science was "to discover and illuminate truth," which included revealing the beauty of nature.[10] She was not ashamed of her emotional response to the forces of nature. Her broad view as an ecologist was augmented by a spiritual closeness she felt to the individual creatures about whom she wrote.

Before she published her book *Silent Spring* in 1962, no one had perceived the permanent damage to the earth that industrialized activity almost invariably causes. Carson saw the wholesale killing of birds and harmless insects by toxic chemicals, and worked to save the beauty of the living world. She recognized that the human and natural environments are interpenetrating, and that modern civilization is literally poisoning the human habitat. It required 350 pages of text supported by fifty-five references for Carson to show how the indiscriminate spraying of pesticides like DDT caused widespread permanent biological destruction. And because the damage to the environment was so unexpected, Carson had to present and explain many of the biological and ecological principles that accounted for the damage.

With *Silent Spring*, Carson forced upon readers a reconstruction of their world. The immediate effect of the book was the banning of DDT as an insecticide; the long-term effect, as one editor put it, was to change the world: "A few thousand words from her, and the world took a new direction."[11] When *Silent Spring* first appeared, the idea that modern technology could annihilate us by irretrievably destroying our habitat was a revolutionary thought—one that provoked debate and discussion throughout society and was greeted with alarm and controversy in government and industry. Although the argument between environmentalists, government, and industry continues today, it is no longer a question of whether industrialization and technology cause damage to the environment, but rather of how much damage is acceptable.

But Rachel Carson was not the first to be troubled about the impact of humans on the environment. In the twelfth century, Hildegard of Bingen was also concerned about pollution. She classified the fish in the Rhine in order to study the repercussions of dumping wastes in the river. Hildegard believed that the ultimate sin is ecological, causing a rupture in the cosmos, a rupture in relationships—for in injuring creation's interdependent balance she saw us destroying all life, including our own. Hildegard shouts to us across eight centuries of silence: "The earth should not be injured. The earth should not be destroyed. Without nature, humankind cannot survive."[12]

In spite of Hildegard's and Rachel Carson's warnings that all of nature is interdependent, the masculine tendency to compartmentalize still prevails. While we now know intellectually that everything is interconnected, we often act as though we are isolated from activities that do not affect our daily lives. The damage to the environment continues, excused by short-term justifications: old-growth forests in Washington State are clear-cut to prevent loss of jobs in the timber industry, equatorial rain forests are disappearing at a rate of a hundred acres per minute to plant fields to feed the hungry, toxic wastes are accepted on Indian reservations because they create jobs, more and more species such as the mountain gorilla are threatened with extinction because humans need more land, oil slicks are tolerated as the price of our consumer society. Although the short-term needs are pressingly real, sacrificing the environment to solve them only delays the impending crisis. In many cases more serious and larger-scale problems result from ignoring the fundamental environmental questions and imposing quick fixes that do not address the deeper issues.

The science of ecology brings to our attention the interconnectedness of the web of life. Because of this interdependence, whenever one species is eliminated, the whole system is affected. Ecosystems tend toward maturity, or stability, and in doing so they pass from a less complex to a more complex state. Whenever an ecosystem is used, and that exploitation is maintained—such as when a pond is kept clear of encroaching plants or a woodland is grazed by domestic cattle—the maturity of the ecosystem is effectively postponed. Many plants and animals require the centuries-long continuity of an ancient forest to fill their role in the complex, delicate web of life that sustains the forest. Even the dead parts of the forest—the decaying logs, twigs, and needles on the forest floor—are involved in recycling and provide habitats for wildlife. And the dead parts are almost completely missing in a young managed forest.

Those without the vision of wholeness can offer only impartial solutions to environmental problems, since they are still limited by compartmentalized thinking. For example, while the Endangered

Species Act helps to prevent extinction of individual species such as the spotted owl (an indicator species for the old growth forest ecosystem) it does not address the larger questions of biodiversity and the welfare of other wildlife species. Rather than preserve old growth forests and retrain those who would lose their jobs, the Interior Department proposes solutions ranging from egg transplants and captive breeding programs, to feeding the owls in the wild. When one species is threatened, it indicates that an entire ecosystem is threatened. Perhaps once we understand the interconnectedness of all things at a gut level, then we will create an Endangered Ecosystem Act to replace the Endangered Species Act that protects a habitat only after a species reaches the brink of extinction. Scientists and citizens aligned with conservation groups such as the Sierra Club, the Nature Conservancy, the Wilderness Society, and Greenpeace have been the primary advocates of legislation to protect our environment.

The new field of "ecological economics" is an example of the broadening vision needed to take us into the future. Participants in this area are trying to translate values like preserving natural resources into the bottom-line language of business—money. Formed in 1988, the International Society for Ecological Economics held its first meeting in May 1990. It has its own journal, *Ecological Economics*. Designed to provide a bridge between natural sciences and economics, this society brings together people interested in revising our concepts of economics so that they are consistent with physical and biological laws. Members of the society believe there are limits to growth, and some think the limits have already been reached. Many see the time as ripe for an economics based not on growth, but on "sustainability." According to World Bank economist Herman Daly, this calls for replacing the old paradigm of the economy as a self-contained system with one that treats it as a subset of the biophysical system. He says that currently "there is no point of contact between the macroeconomics and the environment." Leading economics textbooks do not even contain entries on such topics as natural resources, pollution, and depletion, because most economists treat environmental functions as "externalities."[13]

One theme that emerged from the society's first meeting was

234

"intergenerational equity"—making decisions that will not compromise life for future people. Traditionally, economists have assumed that all resources belonged to the current generation. Shortsightedly valuing the present over the future means, for example, that a slow-growing timber stand can't compete with a dam that provides immediate payoffs. Mathematician Colin Clark of the University of British Columbia said, "much of apparent economic growth may, in fact, be an illusion based on a failure to account for reduction in natural capital."[14]

Today, the price of denying the Feminine, of denying the interconnectedness of everything, has resulted in endangered ecosystems, environmental pollution, acid rain, global warming, and the threat of nuclear disaster. The vision of wholeness that is fundamental to ecology can be one of the unifying forces that can help heal the Earth. As Soviet astronaut Yuri Artyukhin said when he saw the Earth from space, "It isn't important in which sea or lake you observe a slick of pollution. . . . You are standing guard over the whole of our Earth."[15]

MULTIDISCIPLINARY PERSPECTIVES AND INTEGRATIVE THINKING

Ecology gives us a model for bringing together people with expertise in many different areas. This multidisciplinary perspective can enrich almost any area of science. As Davida Teller, a leading infant-vision expert, puts it:

> Visual science is an extremely interdisciplinary field. It includes people from psychology, optics, physics, physiology, anatomy, biophysics, engineering, optometry, ophthalmology, and other disciplines as well. My goal is to take data and concepts that would have originated in many different disciplines and to try to weave them together into a single whole. That's my point of view towards the work, and that's what I try to teach students. I think that my fondness for this particular field, and my teaching within it are very much influenced by my fondness for integrating ideas from all sources.[16]

As we talked, Teller looked out the window of her home on Lake Union in Seattle. She said she always feels a sense of calmness come over her as she watches the water and the ducks. A professor of psychology at the University of Washington, Teller has joint appointments in the Physiology and Biophysics Department in the School of Medicine, and in the Women Studies Program. She directs a research group that ranges in size from five to fifteen people.

After ten years of studying vision in adults, Teller shifted to a knottier question: How does vision develop in infants? Developmental questions have been one of those areas that have been notoriously difficult to penetrate using an unintegrated, piecemeal approach. Interestingly, this shift in her research occurred after she had two children of her own, which stirred up questions about how infants see the world. She found that the empathy she developed in relating to her own children made a tremendous contribution to her ability to successfully coax two- to three-month-old infants into revealing what they saw. As a pioneer in the field, she developed a method by which infants could tell her what they see using their natural tendency to stare at things. This technique is now being used in clinics to test whether an infant's vision is developing normally. Like many other women, Teller brings all of her life experience to her work.

Several years ago I was hospitalized for acute abdominal pain. By the time the x-ray and ultrasound diagnostics were done it was early evening. My doctor consulted with two specialists: the gynecologist was convinced that I had pelvic inflammatory disease; the gastroenterologist thought I had appendicitis. Each physician saw my body only in terms of his own specialty. Both wanted to operate on me immediately. They left me in my hospital room with a form to sign that gave them permission to remove my appendix, fallopian tubes, ovaries, uterus—and anything else in the vicinity that looked questionable. When they returned to my room and said they would not be able to operate until late that night because "things had not been going well in the operating theater and they were behind," I opted for the alternative they offered as an afterthought: to take IV antibiotics and see how I felt in the morning. Fortunately I felt better and did not need surgery. The cause of my pain remains a mystery. This experience brought home to me the degree of fragmentation of our healthcare system and demonstrated how doctors' specialized training colors their interpretation of symptoms. When the experts and specialists disagree, who decides on the course of action? Who can we turn to who understands the interactions between all the complex systems of our bodies, minds, and spirits?

At the national level, the compartmentalization of healthcare is illustrated by the fact that the National Institute of Mental Health (NIMH) and the National Institutes of Health (NIH) are in separate administrative agencies. While institutes covering the eye, the heart and lung, arthritis and musculoskeletal diseases, diabetes and digestive diseases, and allergy and infectious diseases all fall within the umbrella of the NIH, the NIMH is a totally separate agency within the Public Health Service. Such a division vividly reflects the prevailing mind/body split in medicine. Medical science is, however, one of the areas moving toward a more holistic approach. The American Holistic Medical Association (AHMA), formed in 1978, now has 600 members. Although their membership is only 0.25 percent that of the American Medical Association (with 241,000 members), the AHMA is leading the movement to expand the drug-oriented approach of "practicing medicine" to encompass the broader concept of "healing." This pioneering group stresses the integration of physical, mental, emotional, and spiritual concerns with environmental harmony. They speak of healing in the true meaning of the word, which is wholeness.

The vision of wholeness can spring from the intuitive function. Biochemist Patricia Thomas is a highly intuitive scientist who can integrate data from her experiments together with data from other people's papers. She often finds that very diverse papers, which seem to say opposite things, actually say the same thing, and are understandable when seen in terms of the whole system—or through the lens of an integrative new theory. Unfortunately, she has had difficulty finding funding for this combination of experimental and theoretical work. Journals frequently return her papers, asking her to address only a narrow area in each paper. She is concerned that her field is so reductionistic, that more descriptive, integrative work does not get funded. Even reviews of the field only present an accumulation of information, rather than trying to create an integrated story of the whole system. She feels frustrated and believes science would benefit from supporting her intuitive type of "thinking science" that pulls together the threads of many researchers. "The integrative approach is not supported in science today because you're not doing an experimental science. It's a thinking science. There's almost no room for a theoretical science at all—for using other people's data and making these missing links."[17]

One of the pitfalls of science is to follow an experimental path to the point of obscurity, asking smaller and smaller questions so that finally the relevance of the work becomes lost in a morass of data. Keeping one eye on the big picture prevents researchers from getting lost in the myriad of detail, focuses them on the relevant experiments, and helps them relate their work to the needs of society.

What serves to remind scientists of the "big picture" is an individual thing. For example, zoologist Aimee Bakken is happiest when she is doing things in the lab, especially when she is looking at cells through the microscope. When she spends more than three days away from the lab bench, she starts getting restless and the lab draws her back again. She is currently studying chromosome structure and function in frog oocytes (eggs), and her work ranges from looking at isolated genetic material using molecular biological techniques, to watching through a microscope as eggs are fertilized and the embryo begins to divide. Despite the pressures of obtaining funding through grant applications, Bakken feels that writing grants helps her relate her work to the larger questions:

> As much as I might complain about having to write grant applications, they serve a very good function in terms of forcing you to step back, look at what you're doing, and see where it fits into the big picture. Because on the day-to-day basis, just doing one more RNA [a nucleic acid related to DNA that is involved in synthesizing proteins] isolation after another, you can lose sight of the big questions. It also gets very, very boring.
>
> Whatever I do, the questions I ask always relate back to the biology of the oocytes. I've never been really excited about studying RNA protein chemistry. I have to consider that in terms of what may be happening to the RNA molecule when it moves from the nucleus out to the cytoplasm. I feel really strongly that it's important to have a broad background in biology so you can see the big picture and constantly refer back to it.[18]

Bakken thrives on her work. She laughingly says that she would be perfectly happy coming back as a biologist in her next life. She derives the most pleasure from studying whole systems, in her case oocytes, rather than invisible biochemicals in a test tube:

> I can force myself to work with molecules I can't see in the test tube, but I don't enjoy it as much. I get a tremendous sense of joy from being able to see something *alive*, like when an oocyte pinches in two—it's a miraculous process! I've been doing some experiments lately with antibodies and looking at different proteins that the RNA may be bound to, different compartments in the cells, different times during oogenesis [formation of the oocyte], but I keep coming back to asking what this means in terms of the whole oocyte and what it's going to contribute to the embryo once fertilization occurs.[19]

WHOLENESS IN CHAOS

The new science of chaos also speaks to the concept of wholeness. It offers a way of seeing order and pattern where formerly only the random, the erratic, and the unpredictable had been observed. Like ecology, chaos science is a multidisciplinary field. In fact, pieces of the puzzle of chaos were discovered in such diverse areas of science—mathematics, climatology, population biology, physiology, physics, astronomy, economics—that it took decades to connect the pieces and weave them together. Today, the revelations of chaos science are slowly having an impact on the way researchers in all fields view the world.

At the frontiers of mathematics, the new science of chaos shatters, once and for all, the reductionistic notion that a system can be understood by breaking it down and studying each piece. Chaos theory demonstrates mathematically that a system's complicated behavior can emerge as a consequence of simple, nonlinear interactions of only a few components. The science of chaos provides a conceptual framework within which to describe qualitative behavior of systems as diverse as clouds, electrical noise, and the beating of the heart. The interaction of components on one scale can lead to complex global behavior on a larger scale that, in general, cannot be deduced from knowledge of the individual components. In a nonlinear system the whole is much more than the sum of its parts,

239

and it cannot be reduced or analyzed in terms of simple subunits acting together. The resulting properties can often be unexpected, complicated, and mathematically intractable. Edgar Mitchell sees chaos theory as a major force toward holism in science:

> Chaos theory breaks away from reductionism and linear thinking where you could just add solutions and get to an answer. Chaos theory says that the universe is nonlinear. You have to study it in its whole aspect. Computers have helped us do this because there is no equation you can write and get an answer. It's trial and error. You have to run the solution and see where it goes. You have to run down the road and see where it takes you. If you change the initial conditions a little bit and run down the road again, it may take you somewhere totally different. That sort of approach is helping scientists to understand the holistic—or nonlinear—nature behind all processes in the universe. Nonlinear things are complicated, so we've not looked at those before. Now we're forced to. And that means a holistic approach—and studying "chaos." Before, science always presumed that the organized universe began as a random event, a happenstance, an accident due to random noise in the universe. But we discover in chaos theory that the deeper and deeper you look at so-called random motion, the more you see that there is always a pattern at a deep level. It's not random at all. There is order within randomness.[20]

Physicists know all the equations that describe the normal behavior of things such as rolling streams, swinging pendulums, and electronic oscillations. The mechanics seem perfectly well understood. These systems become too complicated for analysis, however, when they go through a transition on the way to chaos, such as from a gently flowing stream to a raging torrent. Understanding the global, long-term behavior of these systems seems impossible. Mathematician Mitchell Feigenbaum, one of the pioneers in the world of chaos, demonstrated with his calculator that well-understood equations were irrelevant when it came to systems in transition to chaos. He observed:

> The whole tradition of physics is that you isolate the mechanisms and then all the rest flows. That's completely falling apart. Here you know the right equations but they're just not helpful. You add up all the microscopic pieces and you find that you cannot extend them to the

long term. They're not what's important in the problem. It completely changes what it means to *know* something.[21]

When a system moves toward disorder, it begins by splitting into two streams, the way smoke rises from a cigarette. Then each of those two streams splits again, forming four streams. This splitting continues to cascade in a process called "period doubling." Feigenbaum discovered, embedded in these systems on the way to disorder, a universal number, a constant as fundamental as π, representing the ratio in the scale of transition points during the doubling process. This constant, 4.4492016090, predicts *when* the splits will occur. He found that when a system works on itself again and again, it exhibits change at precisely the same point along the scale. Such systems are self-referential—that is, the behavior at one level, or scale, guides the behavior of another hidden inside of it. This universality, where different systems behave identically, was hard for physicists to accept. But the constant 4.4492016090 has been found in such diverse systems as animal populations, electrical circuitry, and business cycles. This indicates that our perception of the structures of systems depends on the way we look at them. If we look at them in a certain way, we see they repeat themselves on different scales. Some quality is preserved while everything else changes.

Ironically, because Feigenbaum's discovery in 1976 encompassed both mathematics and physics, it took him two years to find a journal to publish his work. Even though this paper became a turning point for mathematics, one editor still argues that the paper was unsuited to his journal's audience of applied mathematicians. This is a good example of what we miss by compartmentalization.

The interconnectedness of all things is also described by the "butterfly effect" of chaos science, which says that a butterfly flapping its wings today in Tokyo can transform storm systems next month in New York. More technically termed "sensitivity to initial conditions," this describes the dramatic effect that small changes can have on large systems through underlying webs of relationship.

The butterfly effect received its name from U.S. meteorologist Edward Lorenz in the early 1960s. He was using a computer to solve nonlinear equations that simulated a simplified model of a weather system. When he repeated one forecast in order to check some details, he did so using numbers that had been rounded off to three decimal places instead of the six he had used in the previous run. He was shocked to discover that the new forecast was totally different. The small difference between the three- and six-decimal-place numbers had been tremendously magnified by iteration, repeatedly solving the same equation, each time using the answers from the previous calculation—like compounding interest in a savings account.

When Lorenz realized that such small changes in the conditions of initial temperature and air pressure could result in such vastly different systems, he concluded that "the flapping of a butterfly's wing can change the weather." The iterative nature of the nonlinear equations represents the interconnected nature of dynamical systems. In such systems, no amount of additional detail will help perfect prediction. These systems are so sensitive that the smallest detail can affect them. This vast sensitivity to interconnectedness materializes as unpredictability, chaos. Whenever scientists try to separate and measure dynamical systems as if they were composed of parts, they must round off the data at some point. Since there will always be "missing information," dynamical systems such as weather will never be totally predictable.

By mathematically demonstrating the interconnectedness of natural systems, and showing the way these nonlinear systems amplify small changes, chaos theory shatters the deterministic world view. It says that there *is* free will. What one person does in the world *can* make a difference.

QUANTUM RELATEDNESS

The analytical masculine philosophy espoused by the Royal Society of London led to the concept that the universe is like a clock, each independent part interacting with another part by the pushing and pulling of gears and rods. The pushing and pulling forces of interaction were not believed to affect the inner nature of the parts. This idea that you can reduce everything to a machine remains the basic approach of most scientists today, in spite of the revelations of relativity and quantum theory. Advocates of the mechanistic program assume that everything can eventually be treated in this way. David Bohm, one of the world's foremost theoretical physicists, says of this model of nature:

> The mechanical model makes nature a means to an end. It implies that nature is there for us to get whatever we want out of it. I say this model is not adequate. I'm not against treating things as parts, but we have to understand what the word *part* means. A part has no meaning except in terms of the whole. The idea of treating everything as only parts may work in the short run, but it doesn't work when you follow it through.[22]

Although quantum theory has great predictive value, the meaning of the theory remains mysterious. In contrast to the solid world of Newtonian physics, the murky quantum world is described by such unsettling concepts as entanglement, uncertainty, potentiality, probability, tunnelling, wave/particle duality, randomness, regeneration, and the influence of the observer. In some ways, quantum theory is like the insane child in the attic whom everyone would rather ignore. Few attempts have been made to relate these concepts to our everyday world, partly because quantum physicists only deal with quantum theory mathematically, without considering what it means beyond that level. Also, science has become so specialized that other scientists have only a vague idea of what quantum mechanical field theorists are doing. Bohm, however, found that quantum theory provided a theoretical basis for the "interconnectedness of everything" described by Hildegard and others. Curiously, his theories of wholeness made him a maverick among his scientific colleagues, some of whom ridiculed him as a scientist "gone over the hill." His credentials were, however, impeccable.

He received his Ph.D. in physics from the University of California at Berkeley, the last graduate student to study with J. Robert Oppenheimer before the latter went to Los Alamos to direct the atomic bomb project. While teaching at Princeton he wrote the textbook *Quantum Theory* (1951). Praised by Albert Einstein, this book has become a classic in the field of quantum mechanics. Also widely used in universities are his books on the philosophical meaning of quantum theory and relativity: *Causality and Chance in Modern*

Physics (1957) and *The Special Theory of Relativity* (1961). More recently he has written *Wholeness and the Implicate Order* (1980), and *Science, Order, and Creativity* (1987). He earned the admiration of his colleagues through his work on plasma in magnetic fields, his extension of plasma theory to metals, and his contributions to the design of instruments such as the cyclotron and synchro-cyclotron. He even has two phenomena named after him, "Bohm-diffusion" and the "Bohm-Aharonov effect." Until his death in 1992, he was emeritus professor of theoretical physics at the Birbeck College, University of London.

John Briggs, a popularizer of Bohm's ideas, described Bohm as "a pale modest man given to wearing crew-neck sweaters and professorial tweed jackets and to sitting for long periods in apparent passivity while he listens to the conversation swirling around him. But when the topic touches upon science or transformation, his aspect changes. His voice rises, his hands move, and his fingers tremble like feelers leading the way though subtle turns of logic that are shocking in their clarity. During these moments Bohm's utter absorption in his science and philosophy of wholeness becomes charismatic."[23]

His studies over the last fifty years into the implications of quantum theory brought him to the conclusion that the world of the atom is seamlessly interwoven and should not be viewed as a mere collection of unrelated parts. He proposed that a hidden order is at work beneath the seeming chaos and lack of continuity of the individual particles of matter described by quantum mechanics. He termed this hidden dimension the "implicate order," the source of all the visible (explicate) matter of our space-time universe. While modern physics tries to understand the whole reductively by beginning with the most elementary parts, Bohm proposed a postmodern physics that begins with the whole.

Bohm used the metaphor of the hologram to give a static image of his "implicate order" theory of wholeness. A hologram is a photographic record made by laser light. Shining another laser beam through the plate creates a three-dimensional projection. Unlike a photograph, the image stored on the photographic plate does not resemble the original object. Instead, each portion of the plate contains information from the whole of the object. If a piece of the holographic plate is broken off and illuminated with a laser beam, the whole image is still projected, only it is fuzzier. The hologram suggests a new understanding of the universe, in which information about the whole is enfolded in each part, and in which the various objects of the world result from the unfolding of this information. But holograms do not capture the dynamic movement that Bohm saw as basic to the overall implicate order in the universe, where each flowing "part" carries within it an implicit image of the continuously changing and unfolding whole.

Bohm's universe is a process of movement, continuously unfolding and enfolding from a seamless whole. To help us visualize this "holomovement," Bohm uses a device consisting of two concentric glass cylinders, the inner one fixed and the outer one capable of rotation. Between the two cylinders is a viscous liquid such as glycerin. When a drop of insoluble ink is placed in the fluid and the outer cylinder is slowly rotated, then the drop of dye is drawn out into a thread. Eventually, the thread becomes so thin as to be invisible. Now if the outer cylinder is slowly turned backward, the fluid retraces its steps exactly, and suddenly the drop of ink is reformed and visible again. The ink had been enfolded into the glycerin, and was unfolded again by the reverse turning. If a second drop of ink is added after enfolding the first drop, and then the cylinder is turned backward, the drops will appear and disappear at different times. Although it would look as if a particle were crossing the space, the whole system is, in fact, always involved. To Bohm, all parts of the universe are fundamentally interconnected, forming an unbroken, flowing whole:

> Classical physics says that reality is actually little particles that separate the world into its independent elements. Now I'm proposing the reverse, that the fundamental reality is the enfoldment and unfoldment, and these particles are abstractions from that. We could picture the electron not as a particle that exists continuously but as something coming in and going out and then coming in again. If these various condensations are close together, they approximate a track. The electron itself can never be separated from the whole of space, which is its ground.[24]

Bohm proposed that the whole is enriched by introducing diversity, the different aspects of different individual parts, and achieving the "unity of diversity." To him, the real music of the universe is the *unity* of unity and diversity, or the wholeness of the whole and the part. This is another way of looking at the order (unity) within chaos (diversity) described in chapter 6, where order and chaos are intertwined and balanced. Ecologists also speak of the value of diversity as they work toward preserving the diversity of species and ecosystems.

Whereas the mechanistic picture regards discrete objects as the primary reality, and the enfolding and unfolding of organisms as secondary phenomena, Bohm suggested that the holomovement, the unbroken movements of enfolding and unfolding, is the primary reality. An essential part of his proposal is that the whole universe is actively enfolded to some degree in each of the parts. Whereas in the mechanistic view (the explicate order) the parts are only externally related to each other, Bohm's interpretation is one of internal relatedness. In his technical writings he shows the mathematical laws of quantum theory can be understood as describing the holomovement, in which the whole is enfolded in each region, and the region is unfolded into the whole.

In relating his interpretation of quantum theory to the origin of the universe, Bohm said:

> Imagine an infinite sea of energy filling empty space, with waves moving around in there, occasionally coming together and producing an intense pulse. Let's say one particular pulse comes together and expands, creating our universe of space-time and matter. But there could well be other such pulses. To us, that pulse looks like a big bang; in the greater context, it's a little ripple. Everything emerges by unfoldment from the holomovement, then enfolds back into the implicate order. I call the enfolding process "implicating," and the unfolding "explicating." The implicate and explicate together are a flowing, undivided wholeness. Every part of the universe is related to every other part but in different degrees.[25]

> Matter is like a small ripple on this tremendous ocean of energy, having some relative stability and being manifest.[26]

INTEGRATED LIVES

Truly great scientists such as Hildegard, Einstein, Feynman, and Bohm do not compartmentalize their lives into scientific and personal categories. They do not dismiss the implications of their work with, "That is a philosophical question, that's a question for metaphysics." On the contrary, their work affects their personal philosophical and spiritual view of the world. And they realize that their psychological development and inner life make a great impact on the nature of their work. Theoretical physicist Wolfgang Pauli, who received the Nobel prize for his formulation of the "exclusion principle" of quantum theory, frequently met with the psychologist C. G. Jung for a fruitful exchange of ideas on the relations between nuclear physics and psychology. Einstein wrote, "I maintain that the cosmic religious feeling is the strongest and noblest motive for scientific research."[27]

Bohm exchanged ideas extensively with Krishnamurti and also with His Holiness the Dalai Lama of Tibet. Unlike most scientists who keep their spiritual ideas private, he published his metaphysical discussions—those with Krishnamurti are published in the books *Truth and Actuality* (1978) and *The Ending of Time* (1983), and those with the Dalai Lama in Renée Weber's *Dialogues with Scientists and Sages* (1986).

Ideas about wholeness are not just abstract philosophical, metaphysical concepts. They determine the way we look at the world. They affect the way we diagnose and treat disease. Bohm linked his theories of wholeness to our daily lives:

> We don't think our fragmentary approach to reality is a problem because most of us have an unconscious metaphysical assumption that nature is made of separate parts. The eye is a part, the ear is another part, and these parts interact. I'm suggesting that reality isn't that way. If you have something wrong with your eyes, our assumption is that the trouble originates in that part. But it may not. It may originate in the whole of the body, in the mind, or in society.
>
> For example, the problem may be stress or pollution. The society we have created will cause a breakdown in all kinds of parts. You may temporarily repair the parts, but it's like pouring in pollution upstream at the same time that you're trying to remove bits of it downstream.
>
> Pollution itself is typical of the fragmentary approach. Perhaps it's the most striking example. Everybody doing his thing, making his bit of money, and producing his product, and therefore adding his little bit to the pollution. Because the world is finite, all these little bits affect each

other, so that the soil and air are poisoned, fish die, and the climate is changed.[28]

Bohm called for the creation of a postmodern science that does not separate matter and consciousness, so that facts, meaning, and value equally inform science. Science would then have an inherent morality, and truth and virtue would not be kept apart as they currently are in science. His proposal runs contrary to the prevailing view that science should be a morally neutral way of manipulating nature, either for good or evil, according to the choices of those who apply it:

> If we can obtain an intuitive and imaginative feeling of the whole world as constituting an implicate order that is also enfolded in us, we will sense ourselves to be one with this world. We will no longer be satisfied merely to manipulate it technically to our supposed advantage, but we will feel genuine love for it. We will want to care for it, as we would for anyone who is close to us and there enfolded in us as an inseparable part.
>
> Because we are enfolded inseparably in the world, with no ultimate division between matter and consciousness, meaning and value are as much integral aspects of the world as they are of us. If science is carried out with an amoral attitude, the world will ultimately respond to science in a destructive way.[29]

While overemphasizing the reductionistic approach, scientists have paid little heed to the holistic perspective. In doing so, we miss the opportunity to solve previously intractable problems. Ultimately, we could make the planet uninhabitable by continuing this one-sided approach. Just as individuals must integrate the Masculine and Feminine in their journey toward wholeness, so must science embrace the Feminine in order to more fully serve the planet.

In the next chapter we will consider the ramifications of a science in which wholeness is the dominant theme.

The Social Responsibility
of Science

A researcher involved in the Apollo moon project observed, "The real scientific contributions are very commonly made by a single-minded, highly biased individual with a devil-may-care, the hell-with-them-all attitude."[1] This type of person delights in solving challenging problems but neglects to consider the consequences of the solution. Taken to the extreme, this separation of thinking and feeling allowed Nazi scientists to reduce Jews to objects of an experiment in order to study the limits of human pain. Such an attitude contributes to the alienation between humans and nature, and between science and society.

In contrast to the masculine emphasis on separation and autonomy, the relatedness of the Feminine weaves together the individual and community, the situation and the environment, the research and its consequences. As discussed in chapter 3, Carol Gilligan's study of women's moral development demonstrated that an ethic of care and responsibility may be more natural to women than a hierarchical, rule-dominated ethic of rights.

In questioning Apollo scientists about the role of morality in science, Ian Mitroff found that the respondents conceived of scientific morality, if they granted it a place at all, solely within the limited context of stealing ideas or data. One geologist said, "In most of my experience, scientific problems deal with morality very little. I can't think of a specific case where any of this has anything to do with morality because this science deals mostly with the inanimate part of the living world."[2] Another scientist said, "This [morality] has to do with people. This has very little to do with science." Few had grappled with concepts of ethical or moral judgments, or had considered the complex issues involved in the moon program.

Part of the feminine principle of relatedness is to think contextually, to think about how each piece of new knowledge about the world relates to the whole—to see science in the context of the rest the world, rather than locked in the ivory tower of abstraction. In this chapter we will explore a number of questions. To what extent does answering a question impose on the scientist the burden of responsibility for social use of the knowledge? What would happen if scientists refused to participate in potentially socially destructive projects such as nuclear or biological warfare? Does love and respect for nature motivate researchers to look for more humane ways to conduct their research? When does the cost of an experiment, the cost of knowing this one new piece of information, become too great?

In my scientific training I never took a course dealing with bioethics or the social responsibilities of scientists. None were offered. I do not even recall discussing these issues either in classes or informally. I accepted the assumption that all knowledge was equally valuable—that the knowledge itself was neither good nor evil, it was just neutral information. It was up to society whether to apply that knowledge for constructive or destructive purposes.

Yet during a summer job as a technician in a pharmacology laboratory I began to question the cost of that knowledge. In my job I tested the interaction of various drugs. Every day I had to kill a guinea pig to get an inch of its small intestine to test the drugs on. I saved my favorite, black and white "Patches," until the last day, and felt nauseous when I killed him. When it came time to do my own research, I knew I couldn't stomach doing work that involved so much waste of life. I avoided it whenever I could—I used slaughterhouse material or coordinated with other researchers who were "sacrificing" animals for their own experiments. The waste of animals that are killed thoughtlessly—often with little justification—still distresses me.

I know other researchers bristle at the suggestion of anything that limits their freedom to pursue a scientific problem or publish a result. After publication, they are off hunting for the next piece of the puzzle. The norm of impartiality confines them to their area of expertise and compartmentalizes responsibility for knowledge. Scientific doctrine says that "impartial scientists concern themselves only with the production of new knowledge and not with the consequences of its use." For example, J. Robert Oppenheimer, Director of the Los Alamos Scientific Laboratory during development of the atomic bomb, drew a line between pure and applied science:

> The scientist is not responsible for the laws of nature, but it is a scientist's job to find out how these laws operate. It is the scientist's job to find ways in which these laws can serve the human will. However, it is not the scientist's job to determine whether a hydrogen bomb should be used. This responsibility rests with the American people and their chosen representatives.[3]

This way of compartmentalizing pure versus applied science, or basic science versus technology, allows scientists to abdicate responsibility for the consequences of their research. Saying that scientific knowledge is neutral allows the unimpeded pursuit of intellectual interests.

Before explosion of the first atomic bomb, the long-term consequences of radiation on humans and the environment were unknown. In the crucible of wartime, questions of such magnitude as "Would it ignite the atmosphere in nuclear reactions and end us all?" were set aside in the rush to create the ultimate weapon. After Hiroshima, Oppenheimer struggled to resolve the moral problems that arise from scientific discovery, and spent the last years of his life thinking about the relationship between science and society. He confessed, "In some sort of crude sense, which no vulgarity, no humor, no overstatement can quite extinguish, the physicists have known sin; and this is a knowledge which they cannot lose." While most of the scientists at Los Alamos felt they had simply done a necessary job to help win the war, physicist Freeman Dyson agreed with Oppenheimer.

> The sin of the physicists at Los Alamos did not lie in their having built a lethal weapon. To have built the bomb, when their country was

engaged in a desperate war against Hitler's Germany, was morally justifiable. But they did not just build the bomb. They enjoyed building it. They had the best time of their lives while building it. That, I believe, is what Oppy had in mind when he said they had sinned. And he was right. . . . Los Alamos had been for them a great lark.[4]

Oppenheimer served from 1947 to 1952 as chairman of the General Advisory Committee of the Atomic Energy Commission, which in October 1949 opposed development of the hydrogen bomb. In commemoration of the tenth anniversary of Einstein's death, Oppenheimer said, "Late in his life, in connection with his despair over weapons and war, Einstein said that if he had to live it over again he would be a plumber." Broken over the application of his beautiful theories, Einstein told his friends that scientists were caged birds with the system encouraging them to lay eggs, where the disposition of the eggs became the exclusive prerogative of the people in power.[5] He said, "We scientists who released this immense power have overwhelming responsibility in this world life and death struggle."[6] Other physicists shocked by the consequences of their work joined Einstein in forming the Emergency Committee of Atomic Scientists to advocate finding other ways of settling disputes than with the weapons of war.

A somewhat more dramatic example of individuals refusing to contribute to military objectives is that of Irène and Frédéric Joliot-Curie, who jointly received the Nobel prize in 1935 for synthesis of new radioactive elements. As the Nazi menace in Europe moved closer to France in 1939, the Joliot-Curies ceased publication. Rather than reveal their plans for building a nuclear reactor at a time when its unleashed power would have immediate military applications, they secretly deposited their plans in a sealed envelope in the French Academy of Science. During the Nazi occupation, they devoted themselves to protecting French scientists who were at risk. The plans for the nuclear reactor remained hidden for ten years. Later they were instrumental in the construction of the first French nuclear reactor.[7]

These awesome weapons are developed by individual scientists making individual decisions. They are often lured by challenging intellectual problems and causes such as national defense. Many are pushed to work on projects amply funded by the military budget, while other projects such as developing pollution-free sources of energy go begging. Although several steps removed from the civilian men, women, and children who die as the "collateral damage" resulting from their research, the scientist forms a link in the chain. This chain can be broken by individual scientists deciding not to participate. Solid-state physicist Sara Solla at Bell Labs is one who refuses:

> I don't do military work. I think it should not be done, and I think if everybody refuses to do it then it's not done. It's like one of those things, when you say, "Well, what's the difference? If I don't do it then somebody else will do it." That's true, but it is also true that if everybody refuses to do it, then it's not done. So I think one has to stick to that principle.
>
> It's very difficult because sometimes you can be developing something that looks to you very harmless and then eventually has an application that is not what you had predicted and it is outside of your control. Then I think that scientists have to understand that they have moral responsibility for these things as scientists and as citizens, as members of the community. It's a little bit outside of your role as a scientist, your obligation to become politically involved and try to have control over things that people should try to have control over. It's a complicated question. I think the first thing one has to do is to be vigilant. Be aware of this problem and pay attention to it, and ask oneself these questions about the work that one is doing as one goes along. I think one should not be carried by the ingenuity of people who worked for instance in the Manhattan Project, thinking that once it was all done they would have control over it. Looking back it was an incredibly naive way of thinking about it and I think one should learn from that and not be so naive so that one can act before it is too late.[8]

Like Solla and many others of my generation, I struggle to reconcile my love of nature, my curiosity about the world, my desire to make a contribution—with the power and responsibility that knowledge brings. I was brought up to believe that science and technology provide us a way to solve all our problems. After all, a science that puts men on the moon can do anything, given enough time and money. All we need is more knowledge.

That knowledge endows us with power, the power to link peoples of the world through instantaneous communication, the power to develop weapons that forced countries such as Japan and China to open to the West against their will, the power to replant the earth with genetically engineered crops, the power to bring the healing sound of music into the home, the power to vaporize a city such as Hiroshima and poison the earth around Chernobyl, the power to liberate the human spirit from the drudgery of soul-deadening labor. What role does the individual scientist assume in unleashing and wielding this power? Can one person make a difference?

Handling power is arguably one of the biggest challenges human beings face. The old adage "power corrupts" urges caution, yet perhaps causes people to deny that power and abdicate responsibility for handling it. Many hide behind the banner of "pure science"—science for the sake of science, science for the love of knowledge, science for the fun of discovery. Yet basic science and technology are intimately intertwined. Technological advances such as electron microscopes, telescopes, or computers engender further advances in basic science, and are themselves built upon theoretical physics, materials science, and information theory.

As I began to ask myself in my own life "What's enough?" and question my personal goals, I also questioned the goals and assumptions underlying my profession. Do we ever reach the point of having enough knowledge? Enough power? The Feminine helps take these questions out of the abstract and ground them in the context of actual situations. Intuition can offer us a glimpse of the future, an array of possible consequences of new knowledge. And then we need to rely on our feeling function to answer these questions of values and priority.

Other scientists are also stepping back and asking fundamental questions about the value of science, the consequences of progress, and how their work contributes to society and the global economy.

TAKING A STAND

While the atom bomb shocked physicists into thinking about their role in military research, the impact of other basic research may not be so obvious. Agriculture, for example, seems an innocent and peaceful activity. Botanists do experiments to learn

254

about the nature of plants because they are curious about the world. They produce better plants and higher yields. Conventional wisdom tells us that more research in plant biology will solve the problems of polluting the atmosphere and feeding the growing population of the world. Now, however, a few scientists are beginning to question this prevailing assumption of science that the more we know, the better we will be able to manage our environment.

Indiana University biology professor Marti Crouch describes herself as "a born botanist." The daughter of a florist, the granddaughter of a raspberry farmer and mushroom-hunter, Crouch developed an early love for plants. During second grade she made friends with an American Indian woman who took her for walks in the woods and taught her to identify and eat wild plants. As a teenager, she worked at a botanical nature center and met many wonderful botanists who "crawled around on their hands and knees in the forest, ecstatic over finding some little liver-colored orchid."[9] Her love for nature led her to become a botanist herself. It seemed too good to be true that she could get paid for doing what she loved.

Over the years she became well respected for her studies of plant embryo and pollen development. Between 1980 and 1990 she wrote or cowrote twenty-two articles, received almost a million dollars in grants, and obtained tenure at Indiana University. In 1990, at age 38, she concluded a three-year project on rapeseed embryo development funded by a $320,000 National Science Foundation grant.

After obtaining tenure in the biology department, Crouch stepped back from the intense task of proving herself in her field. She began to think about how her work in basic plant science fit into the larger picture of agriculture and the environment. She wondered why the United States government paid her to study plants. To feed starving children? Upon reflection, she doubted it. Her examination of the green revolution led her to conclude that development of higher-yielding plants had not resulted in a decrease in world hunger. While overall agricultural productivity had increased, the primary beneficiaries were those who could afford expensive fertilizers, pesticides, herbicides, and irrigation.

Expansion of export-oriented plantations forced poor farmers to retreat to the cities or to try to eke out a living on marginal land—often slashing forests and displacing wildlife and indigenous people in order to survive themselves.

Instead, Crouch found that those who have benefited the most from the green revolution are oil companies and the multinational corporations who supply the machinery, seeds, fertilizers, herbicides, and pesticides. The bankers who finance roads, dams, and processing plants also profit, leaving a developing country with a burden of debt. Crouch concluded that applications of her work were more likely to profit multinational corporations than the teeming masses of poor and hungry people. In addition, her work supported the systems that contribute to the destruction of rain forests, native cultures, small farmers, and water quality.

Crouch is a woman who cares deeply about the environment. For years she has done her bit by recycling and walking to work. But her reflections brought to light a fundamental conflict between her beliefs and her professional life. Convinced that our civilization is in the midst of a global crisis, she felt she had to dramatically redirect her life. She had to abandon "business as usual."

With a firm commitment to the belief that she can help make the world a better place, Crouch quit experimental science in 1990. She decided to do so publicly and discussed her rationale in an essay to her colleagues in *The Plant Cell* (the official journal of the American Society of Plant Physiologists), "Debating the Responsibilities of Plant Scientists in the Decade of the Environment."[10] Her action sparked a spirited debate among her colleagues. Some have written her long letters expressing their own frustrations and sense of unease. Crouch said, "I've been really surprised at what a broad feeling of unease there is among scientists—particularly students and postdocs, but also some faculty—and how little it is discussed in their training and how anxious they are to talk about it."[11] Numerous groups of graduate students and post doctoral researchers from agricultural schools across the country have invited Crouch to give seminars discussing the social responsibilities of plant scientists. Her seminars provide a forum for considering a wide variety of issues. Crouch has been encouraged by her colleagues' response:

Several hundred people came to my seminar [at Purdue] and there was lively discussion, and much interest and concern about how you change such an entrenched system. We talked about the merits of working within or without the academic system. I've actually been fairly surprised and heartened by the amount of interest.

People are uneasy about many aspects of science: why are we so driven (what's driving us to work so hard, why don't we have any time to do anything else), why are the first products of plant genetic engineering herbicide-tolerant plants, why is the process of what we're doing so environmentally damaging (radioactivity, toxic chemicals, plastic disposables)? A high proportion of the graduate students in U.S. agricultural schools are international students, so they come from a perspective of first-hand experience with Third World agriculture and they say, "We don't want to take back these industrial agricultural methods, but we aren't being taught anything else." There's lots of questioning.[12]

On the other hand, some of Crouch's colleagues simply think she is crazy. Jose Bonner, associate chairman of the biology department at Indiana University, is a molecular geneticist who studies fruit flies and yeast. He reflects the traditional compartmentalized viewpoint in his comment: "All we know now is that she's decided to try to address an important problem, a problem that our department traditionally has not tried to address. It's not supposed to address it. It's a biology department, not a sociology department. That's not what we're trained for."[13]

Although he recognizes the importance of the problem, Bonner felt Crouch's experimental work had tremendously valuable potential. Like most scientists, he believes that knowledge is neutral—that it can be used for good or bad, and that the best a scientist can do is hope that people will use their discoveries "without screwing up." He leaves the ethical decisions to those who apply the knowledge: industry, policy-makers, politicians, and the public. The application of knowledge is simply not his job. He was not trained for that.

Although she has left experimental science, Crouch maintains her university appointment. She continues to teach, read, and attend conferences in the field of agricultural ecology and world hunger. She may apply for grants to analyze effects of science and

257

technology on society, which would involve working with economists, anthropologists, and sociologists.

To some, Crouch's action appears heroic. Yet she says that it was not a difficult decision to make because her sense of identity is not tied up in her work—and molecular biology had become boring because it's "so linear." Her action of conscience does not come as a sacrifice for her, but as an opportunity to make a more meaningful contribution. A lifetime subscriber to *Whole Earth Review* (originally *Co-Evolution Quarterly*), Crouch said:

> When I was twenty years old and reading *Co-Evolution Quarterly* I thought, "I'd like to be able to do this kind of thinking as what I *do*." So the opportunity for me to interact with sociologists and economists and to think much more at the systems level, and to experiment with different ways of teaching is really fun—it's more fun than working in the lab was.[14]

She formed the Bloomington Rainforest Action Group to educate people about their role in rain forest destruction and the value of rain forests, and to suggest ways to preserve rain forests. She also coedits *Forest-Watch Newsletter,* a publication designed to inform citizens about ecological issues.

Although she wants to shake people out of their numbness to global crisis, she is not strident in her approach—nor is she anti-science. She does not condemn researchers or suggest that everyone leave experimental science. She relates her personal experience and explains how she arrived at her conclusions with the hope that others will begin questioning. She then hopes people will reevaluate their research objectives in light of their possibly negative social consequences. She challenges tenured faculty to be more imaginative and less obedient. Since tenure gives them financial security, they have very little to lose—yet they have been the least active. Crouch said:

> I talk to them about how I have an anarchistic view that if people are genuinely working toward the common goal of change, they can do that from several positions. I talk about scientists who are trying to work in a way that integrates the basic research with the result. I talk about the New World Agriculture Group, scientists who work in Central America, directly with farmers cooperatives and research proj-

The Social Responsibility of Science

ects related to their needs. I talk about trying to strengthen incentive systems for local marketing of agriculture and strengthening systems that support subsistence agriculture instead of the other way around. I talk about the idea of changing the funding priorities to favor systems and holistic research.[15]

Some scientists have been inspired by Crouch's action to look deeply at the long-term consequences of their work. Some feel they can better change how science and technology affect the world by remaining in research. For example, June Medford and Hector Flores from the Pennsylvania State University responded to Crouch's challenge by advocating a greater dialogue between scientists and farmers in developing areas in order to better understand the farmers' needs. They also suggest that university scientists begin teaching courses on bioethics and the impact of plant science on society.[16] Steven Smith at the University of Arizona calls for plant biologists to accept their social obligations, "as bothersome as they may be." He wrote, "Failure to do so is not only an inefficient way to conduct science, but is likely to lead to further erosion of the public's trust in us and our products. It is time for plant biologists to demonstrate that we are citizens, as well as scientists."[17]

Crouch exemplifies the feminine principle of relatedness by looking at her work in a larger context and by drawing together knowledge from diverse disciplines. She did not compartmentalize her work, but rather saw the connections between her work, ecology, and the global economy. She is now cooperating with scientists and nonscientists alike toward a common goal of change.

She used her feeling function to judge the value of her research in relationship to the larger issues of world hunger, the destruction of the rain forests, and the disappearance of indigenous cultures. While the thinking function is content to contribute more details to an abstract theory of how the world works, Crouch used her feeling function to evaluate how her research affected other people's lives. In redirecting her professional life she brought it into alignment with her personal values and beliefs. She pierced the illusion of scientific objectivity—of "knowledge for the sake of knowledge"—

and identified the wealthy multinational corporations who lobby for governmental funding of basic plant science in order to further their own interests.

She expresses an appreciation for diversity when she says, "Scientific understanding enriches our lives but is not any more or less valuable than other forms of knowledge." She sees humans as part of an interacting web, where every action "tweaks a thread, causing a complex set of reactions." Rather than manage our environment, she would rather see us use science to humbly and respectfully adapt to what exists. In teaching science to nonmajors, she has created courses such as "The Biology of Food" and "Ecological Investigations of Daily Life" to bring science "home" to students. By making scientific knowledge more accessible, Crouch helps bridge the gap between science and the public.

Crouch's action comes from a desire to nurture the planet rather than dominate it. She takes a long-term approach to her project of changing the rationale of agriculture back toward a decentralized, nonexport-based system of local food production. Through her essays and seminars, Crouch has touched the lives of many scientists and students. She responds personally to them and encourages them in their process of redefining their role in science.

Crouch is an example of a scientist asking fundamentally different questions, such as "What would plant science be like in an ecocentric world?" By seeking to develop a strong social as well as scientific conscience, she seeks to renew our concepts of science. Accepting the obligations of social responsibility, and maintaining the perspective that everything is interconnected, requires redefinition of the limits of scientific freedom. While Barbara McClintock exemplifies a more human science based on receptivity and a "feeling for the organism," Crouch believes we need to go a step further.

What concerns me is that Barbara McClintock is held up as an example of someone who approached science in a unique way and was able to discover new and interesting things because of her unique approach. However, the fruits of her work have been used by the same interests as any other scientific information. In fact, her research has been pivotal in developing biotechnology. So if we develop a network of scientists who are doing things differently, but don't also change the link with the application through the dominant system, it seems that we're just helping to maintain the status quo. That's why I decided not to do experimental science anymore rather than simply change what I was doing, because I felt I could change what I was doing to be more in line with how I felt about nature—but if I found anything out, the end result would be the same in terms of societal use.[18]

SERVING SOCIETY

In her book, *In a Different Voice,* Carol Gilligan describes the struggle between care and responsibility as the "progression of relationships toward a maturity of interdependence."[19] As a woman reaches adulthood, Gilligan explains, "the moral dilemma changes from how to exercise one's rights without interfering in the rights of others to 'how to lead a moral life which includes obligations to myself and my family and people in general.' "[20] Applying this to science might lead us to ask, how can individual scientists serve society? This is no small question. Nearly half of the bills that come before the U.S. Congress have a significant scientific or technological component.[21] I believe scientists must play a larger role in creating an educated public that can make intelligent decisions in our increasingly technological society.

Since most scientific knowledge has the potential for both good and bad, it is critical that scientists reflect on the possible negative consequences, to see science in its broader context, and to shoulder the burden of helping society handle it in a responsible way. For example, even though it is very time-consuming and distracting to her research career, Aimee Bakken believes that it is important to advise the public in appropriate use of technology.

> I feel a strong commitment to educate students and the lay public so that they can read stuff in newspapers and magazines about science and make educated judgements about them. During the seventies, I helped organize a recombinant DNA conference for the lay public, and I'm now testifying as a scientific expert in DNA fingerprinting trials for a rape-murder case. I think there is a real need for scientists to stand up and educate lawyers and judges, and try to get them to develop high criteria for the admissibility of DNA fingerprints in court. The couple of companies that are doing the work are sloppy and they don't seem to care. I'm hoping that an educated legal community will pressure these companies into upgrading the quality of their work.[22]

Marsha Landolt, a comparative pathologist in the School of Fisheries at the University of Washington, has been in the field long enough that people ask her to serve on national committees looking at particular issues such as aquaculture. She derives a great deal of enjoyment and satisfaction from speaking as a scientist to shoreline hearing boards, county planning boards, and pollution control boards. Recently, she was asked to serve on a science advisory board for the Great Lakes, looking at their pollution issues.

> I met with the Great Lakes Board for the first time last month. For me, it's a brand-new group of people with different sets of ideas, looking at things in ways I hadn't considered, so I'm learning something, and that makes it fun. It is more of a synthetic look at things, where you take twenty years of knowledge and try to bring that to bear on an issue, where you may need to look at ten different sides of a single issue to try to understand the best path. Again it's a team sort of effort.[23]

Although some may say "that's not really doing science," Landolt is filling a sorely needed role by helping to bridge the gap between scientists and the public. She feels scientists need to play a much more dominant role in technical-social issues. While some scientists are more comfortable sitting in the lab with the door closed, she prefers to be "out on the periphery" interfacing with the public.

> I've worked for the last fifteen years in the area of aquatic toxicology. I *care* what's happening to fish as a result of man's activities. But in spite of that, I find myself increasingly at odds with the tenets of the environmental community. Aquaculture is one manifestation of that, the flap over Alar in apples, some of the nutritional fads that come up all the time—where people are taking a little snippet of science, using it for their own agenda, and the scientific community is not standing up to say, "Hey, wait a minute! This is really being misrepresented." There is a time for scientists to be isolated and to simply do good work, but there is a responsibility for scientists to be out saying, "What is the question? Why is this important?"—explaining to the people who have supported us through their tax dollars what we've learned or provide information for their personal interest. That's a part of science that I'm very comfortable with. I think there is a need for some balance.[24]

Abstract "knowledge for the sake of knowledge" is not enough to motivate some scientists, yet they still feel vulnerable about expressing their desire to do work that is relevant and related to improving the human condition. For some, this involves working with the hope of providing solutions to environmental problems, to cure a disease, or to enhance the quality of life.

As early as she could think, before she even knew what she was talking about, Sigrid Myrdal knew she was going to work on finding a cure for cancer. Now a Senior Scientist in a cancer research division of Bristol-Myers Squibb, she studies growth factors. Her journey has been a long one, requiring a great deal of determination and tenacity. After earning a bachelors degree in chemistry, she interrupted her education to put her husband through graduate school and have three children. When her husband found a job, she obtained a second bachelors degree in biology, a Ph.D. in developmental biology, and then acquired further training through two postdoctoral fellowships. Along the way she trained herself in the biology of normal and malignant cells. In planning her research, her overriding principle is to be able to state the relevance of her work in one sentence.

> My first year in graduate school I decided: yes I did need to use the scientific method as described to me by my professors and reduce things to testable hypotheses. But I had a laboratory in my head, and on the wall of that lab was a huge sign that said, "So what?" I had to look at that sign and make sure that I could answer it, preferably in one sentence. Otherwise I'd get into some picky little thing.[25]

Even after she realized her goal to do cancer research at Bristol-Myers Squibb, her work was interrupted by a two-year hiatus in which she personally confronted the object of her study—cancer, in the form of leukemia. Facing 50 percent odds of living, she underwent a bone-marrow transplant. In the process of recovery she literally had to learn to think again. Now back at the lab bench, she is filled with gratitude that she can do the work she loves, and she wants that work to be related to improving the human condition. She recalled an off-handed comment from her postdoctoral advisor while discussing his work:

> My advisor said, "Well, I certainly don't want to cure a disease" in the most scornful, acid voice, as though that were beneath his dignity. He said it in a way that you knew it was exactly the right thing to say in a

263

biochemistry department—that anybody who did want to cure a disease would be pretty damn naive, fairy-eyed, not doing pure science, not asking the elegant questions.

And I thought, "Well *I* want to cure a disease!" I don't think they had ever conceived of the notion that you can have a scientific curiosity as a director of your questions and a passion that's fired by something else that pushes you to do the research. I didn't want my reputation to be sullied by wanting to cure a disease, so I just kept quiet about my feelings.[26]

While many researchers have very successful careers doing "so what" projects, they waste considerable valuable resources in the form of time, effort, intelligence, and research materials. But who is to judge? Who decides whether a project is a waste of time? Again, these are questions of worth—of value and priority—the province of the feeling function, a quality of the Feminine. Science leaves these decisions to individual researchers, peer review panels, funding agencies, journal editors, and reviewers of research papers—all gatekeepers of knowledge.

GATEKEEPERS OF KNOWLEDGE

One of the greatest sins that can be commited by a Sufi teacher is to impart knowledge to a student before that student is ready. The Jewish tradition has severe laws against people who speak about the inner secrets to those who are not prepared to receive them. Celtic warriors first had to become poets before they could go into battle. These traditions learned from experience that powerful knowledge acquired too soon can overload a person's ego. Like a child playing with fire, the student may lack the psychological maturity and discipline to handle the power appropriately. Knowledge and power can go to our heads, inflating our own sense of self-worth and entitlement, leading to arrogance and hubris. To truly serve ourselves and others well, knowledge must be balanced by love and compassion, channeled through the heart, and evaluated by the feeling function.

Certainly, those with knowledge and power have withheld them in inappropriate ways at times. But the concept of timely teaching also contains much wisdom. We have been conditioned to think that everyone should be told everything, that access to information is the best prevention of the abuse of knowledge. However, once a technology is possible, we tend to think we know all about it and feel compelled to use it—especially if someone can make money from it. Only later do we discover the dangers in this race for progress, and then it is difficult to stop the momentum. In our fast-paced culture we need to pause long enough to apply discretion and to make individual and collective ethical decisions based on context and timing. Rooted in a sense of connection, the Feminine can help balance "good science" with socially responsible science, short-term "progress" with long-term sustainability.

It is also important how the knowledge is transmitted: who releases the information, how little or how much, the values that accompany it, the political and socioeconomic climate into which it is released, and the level of understanding of the "student" or the public. Some new developments in science and technology may require a moratorium on their application until we, as a society, can debate and consider the social and ethical implications. I am not advocating a secretive science where information is forever withheld. Rather, as gatekeepers of knowledge, scientists could think more deeply and take more responsibility for the long-term applications of their work. Professional associations could also form groups similar to the national Office of Technology Assessment, which tries to understand the dynamics of technology as it evolves in different social contexts.

Science seems to promote a technological imperative like manifest destiny. We rarely consider the values built into new technology, or how applications of science change the social patterns and daily lives of people. Because of this, Einstein saw a disintegration of the social fabric arising from the cold scientific mentality. He said:

> What need is there for responsibility? I believe that the horrifying deterioration in the ethical conduct of people today stands for the mechanization and dehumanization of our lives, a disastrous by-product of the development of the scientific and technical mentality. We are guilty. Man grows cold faster than the planet he inhabits.[27]

In addition to the examples of military research and agricultural research, there are innumerable other areas in which we face ethical

265

choices: computer technology and its impact on the privacy of the individual; who has access to expensive medical technology such as dialysis machines or organ transplants; how we dispose of toxic and nuclear wastes; use of contraceptive technologies for population control; and the application of mind-control technologies. New knowledge about the genetic code may soon make it possible to create "designer genes" to improve health—or alter skin color. Along with new knowledge will arise new opportunities to make individual and collective ethical decisions about the use of scientific knowledge and power. Such decisions are complex and are best addressed by individuals at a high level of psychological maturity. Scientists who take up these ethical issues will be individuals of high capacitance who can hold both sides of a question—suffering the tension of the opposites without falling prone to simple either/or answers—until the creative solution emerges. Struggling with the daily ethical choices, from humane treatment of experimental subjects to application of new knowledge, can stimulate us and challenge us to become more conscious.

The potential danger of continuing a single-minded, devil-may-care, hell-with-them-all approach to science is no less than annihilation of life on the planet. As Celtic warriors were trained in poetry before they were trained to use weapons, so can scientists seek a training of the heart. Let us now embrace the positive contributions of the Feminine that were cast into the shadow at the birth of science. As we have seen, the qualities of feeling, receptivity, subjectivity, multiplicity, nurturing, cooperation, intuition, and relatedness can guide us in dealing with the complex questions arising at the interface of science and society. These are big questions and call on us to use all of our resources—Masculine and Feminine.

LIFTING THE VEIL
The Feminine in Every Scientist

Just as an individual's task at midlife is finding meaning in
life, so the task of science at midlife is a reevaluation of the
meaning of science. It falls to the individuals within the collective of
science to take up this task, to contribute to the evolution of
consciousness, to develop intuition and feeling—to bring the Feminine out of the shadow. Just as molecules in heating water spontaneously form more complex patterns, scientists can begin to
change, receive support through networking, reorganize and amplify themselves through feedback loops, and spontaneously form
new types of organizations. I believe that the institutions of science
will change when we reach a critical threshold of individuals changing. But first it will take intention, will, integrity, and courage to
choose to invest science with new values based on the relatedness of
the Feminine. We will pay a terrible price if we do not bring feeling
and the Feminine out of the shadow and into the light of consciousness.

While the Feminine is emerging in areas such as quantum physics
and chaos science, many scientists cannot face the implications
of this new worldview. Jungian analyst Marie-Louise von Franz
writes of a scientist who had a rigid worldview that illustrates the
extreme logos attitude of the Masculine. This scientist was so
caught up in the mechanical worldview that he ignored the discoveries of modern physics. Whenever von Franz discussed with him
how quantum physics changed the image of matter, he became

emotional. He said that if those things proved to be true, he would have to shoot himself. Because he had been teaching the mechanistic view to students for generations, he should stand by such ideas. If he discovered they were not right, it would be dishonorable for him to go on living.[1]

At some point in a scientist's life, the repressed Feminine breaks through to consciousness and asks, "What does all this mean? What have I contributed? How does it benefit humankind or the planet?" In spite of fame or financial success, if scientists discover their work has been unimportant and meaningless, they suddenly see their lives and themselves as dry and worthless. Because they have dedicated their lives exclusively to science, their sense of identity and self-esteem comes from the value of their work. For a while they oscillate between an inflated sense of superiority and a terrible feeling of emptiness, until the reality of the emptiness exceeds their capacity for denial and rationalization. Long ignored and unintegrated, the dark side of the Feminine engulfs them in despair. Some scientists take their own lives when they realize they are not going to make a difference in a fundamental sense. Others dramatically shift the direction of their careers and their lives.

Once integrated, the Feminine brings deep joy and satisfaction, a sense of unity and connection, a renewal to life. Jungian analyst Ladson Hinton tells of a client at midlife who kept his imaginative vitality compartmentalized and unused. Symbolically, he kept his camera equipment sealed from dust and moisture to the point that he never used it. In his pursuit of perfectionism, he kept each precious object eternally in its place. His constant war against moisture and earth reflected the constriction of his emotional life. His arrogantly high standards labeled other people inferior and kept them distantly in their place. The image of an Egyptian tomb in a dream helped him see the fear and hatred of life implicit in his attitude of perfectionism. Through analysis, he increasingly embraced the Feminine and found his way to a vital and spontaneous life. He turned away from the "necropolis ideal" and began to accept life in its rich diversity and ambiguity.[2]

Insofar as science adheres to the goals of prediction and control, it subscribes to this necropolis ideal. The certain, the simple, and the predictable give us the illusion of safety. In order to conduct our daily lives we require confidence that the world is constructed in a definite way. We need to believe that the sun will come up and that the unexpected will not intrude. Our desire for security is a valuable protective mechanism, but it can also be a prison of denial that keeps us locked in old ways of seeing the world. Anything unpredictable seems dangerous. There is a tendency to make life safer and simpler by owning only one of the dimensions of reality, one of the pair of opposites, and projecting another. Such delusions and defenses help us survive the shame of our finitude and preserve our self-esteem. But they inevitably narrow our perception of ourselves, others, and reality to achieve a comfortable illusion of control. We become blind to the rich dimensions of all there is. The intellectual development of many a scientist comes at the sacrifice of other psychological growth. As Jung said:

> Science as an end . . . leads to a high differentiation and specialization of the particular functions concerned, but also to their detachment from the world and from life as well as to a multiplication of specialized fields which gradually lose all connection with one another. The result is an impoverishment and desiccation not merely in the specialized fields but also in the psyche of every man who has differentiated himself up or sunk down to the specialist level.[3]

The essence of life is change, constant movement, combination, dissolution, recombination—an ambiguous, ongoing alchemical dance of polarities. Until recently, science compartmentalized the chaotic earthly "generation" and "corruption" of nature, holding them at bay by projecting them onto the Feminine. Paradoxically, in the process of trying to tame nature into more predictable behavior, science itself has been a powerful agent of change.

Even more paradoxically, as many people over the past two decades have become disenchanted with science, they have projected their fear of change and fear of the future onto science. Some of these are rational fears. Some stem from a lack of education and understanding. Others come from the separation of science from society, which has been reinforced by separate language and jargon, and the isolation of researchers living out their lives in the lab. Many are so consumed by the business of science that they have little energy left over to participate in community life. But when people become separated from the community, they attract projections. Without the warmth of human contact to dissipate the clouds of projection, researchers become portrayed as mad scientists creating Frankensteinian monsters. In fact, movies and the popular media rarely portray scientists as ordinary people struggling with ordinary problems.

THE SHADOW OF SCIENCE

Before the evolution of consciousness, the shadow is simply the whole unconscious. As the dark, unlived, and repressed side of the ego, the shadow contains all of that within us that we cannot directly know, all the processes at the back of the mind of which we are not aware, all the qualities that are not admitted or accepted because they are incompatible with those we have chosen. Gradually we can sort out three parts to the unconscious: the inferior function (the fourth psychological function to develop, in Jung's typology), the contrasexual elements, and the part of the shadow consisting of brutal emotions and undesirable parts of ourselves we repress. These repressed or undeveloped qualities can be bright and good as well as dark and destructive. For example, a Scrooge may have a fit of unexpected generosity, which later embarrasses him.

When we do not accept a quality of the shadow, then it functions behind our back, leaking out when we least expect it. We say that something "comes over us," that we don't know what "possessed us" to do something, or we find that the "right hand does not know what the left is doing." All people, institutions, and cultures have their shadow. Although we usually cannot see our own shadow, we learn about it through the reaction of onlookers. The collective shadow of a group or a nation is particularly difficult to see because people support each other in their blindness. One of the roles of the feminist critique of science has been, as outsiders, to point out some of the shadow aspects of science.

Another way to catch a glimpse of the shadow is to become aware of particularly strong emotional responses, such as a disproportionate attraction or revulsion to another person—or the aversion of science to psychic phenomena. When we are caught by the collective shadow it can feel like we are "possessed by the devil."

The lynch mob is an extreme example. But these collective demons get us because we have a little bit of them in ourselves. If we have not sufficiently integrated a part of our personal shadow, the collective shadow sneaks in through this door.

In the inception of science, the demands of creating some degree of safety and security by predicting and controlling nature compelled scientists to create order using the thinking function—the function best equipped for finding some order in the chaos of nature. As a rational discipline, the "masculine philosophy" of science became almost completely identified with this function. While sensation became well developed in the hands of experimentalists, feeling and intuition received little attention. Like other repressed or unappreciated qualities, they remained in the unconscious and became entangled with other contents of the shadow.

Today, many of the functions of thinking and sensation can be carried out by machines: computers can manipulate data faster than most human brains; instruments can measure wavelengths of light and sound radiation over a broader range than can be detected by human eyes and ears; gas chromatographs can detect the presence of aromatic substances more precisely than the human nose. But we need intuition to put the data together to form patterns and extrapolate meaning; and we need feeling to determine worth and to assess moral and ethical values. Jungian analyst Irene Claremont de Castillejo writes:

> The deeply buried feminine in us whose concern is the unbroken connection of all growing things is in passionate revolt against the stultifying, life-destroying, anonymous machine of the civilization we have built. She is consumed by an inner rage which is buried in a layer of the unconscious often too deep for us to recognize. She becomes destructive of anything and everything, sometimes violently but often by the subtle passive obstruction.... With more consciousness, feminine anger could be harnessed to a creative end.[4]

In order to reclaim the Feminine from the shadow, we must shine the light of consciousness on its qualities. As we come to recognize the value of those qualities, we can learn to integrate them and express them appropriately as the situation demands. Doing so takes courage, because those around us may discourage us from changing since it means they also have to readapt. To the degree to which we can consciously use all our functions, we are liberated from the tyranny of any one of them.

Jung called the psychological function that lags behind in the process of differentiation the "inferior function." Because it is the least developed, it is still largely unconscious and fused with other contents of the unconscious. It cannot yet be willfully and appropriately exercised. Because it is the function closest to the unconscious, it carries a magical or numinous quality. In the case of science collectively, and of many scientists individually, feeling is the inferior function (in the sense of least developed, not less valuable). To most, feeling is an afterthought in the business of science, if it occurs at all. But a thinker whose feeling is suddenly touched can be overwhelmed by the force of an unexpected emotion. Similarly, when an intuition pops into the mind of the sensate it seems heaven-sent, lit by a glow from the unconscious. Because of the close link to the unconscious, the inferior function is both a gateway to the demonic, as well as the path to renewal in us. This makes it imperative that we develop feeling as a conscious part of science—because it is both the inferior function of science and a repressed quality relegated to the Feminine.

THE DOOR TO EVIL

By recognizing that feeling is the inferior function in science, we can more easily be on guard that evil does not creep in unawares and lead us astray. We must take heed from the example of Nazi Germany and recognize the menace that comes from over-developed thinking and unadapted feeling. Lack of a sound feeling function in Germany, a nation with highly developed thinking, opened the door to the evil of Hitler's regime. In the name of purifying the German nation, Hitler and his close collaborators used their thinking function to efficiently exterminate millions of Jews, Gypsies, and homosexuals. These Nazis had no adequate feeling function in consciousness with which to evaluate the horror of their monstrous actions. So faulty was Goebbels' feeling function that he wept when his canary died, yet he lacked a sound valuing of human life. With their faulty feeling, many Nazis could not consciously and correctly evaluate the evil of their deeds, and so acted with a combination of sentimentality and callous brutality.

Under the banner of national defense, science also crosses the line into evil. While nuclear bombs are arguably used only on the enemy in self-defense, and cause no harm to citizens, our tax dollars have also supported research in the name of "psychological warfare" and "national security" that has damaged hundreds of college students and psychiatric patients—hardly our enemies.

During the 1950s and 1960s the Central Intelligence Agency, the Canadian Defense Research Board, and the Department of Health and Welfare supported mind control experiments.[5] From 1951 to 1956, neurophysiologist Donald O. Hebb, chairman of the psychology department at McGill University, conducted sensory deprivation experiments on student volunteers. As a result of these perceptual isolation experiments, a majority of healthy students developed hallucinations and suffered long-term difficulties in concentration and problem solving. One student became hysterical; another suffered an epileptic seizure. Even after the isolation, many students complained of nightmares and insomnia because of the terrifying nature of the hallucinations. Others complained of dizziness, confusion, nausea, fatigue, and headaches. A majority of the students described the experience as "a form of torture." In spite of these effects, the research continued and expanded. During sessions of perceptual isolation, students were subjected to a series of ninety-minute recorded tapes about ghosts and poltergeists to determine if they would blindly accept data—in other words, brainwashing.

Other brainwashing studies included a combination of sensory deprivation, psychic driving, electroshock, and the use of testosterone on women. Under the code name of "ARTICHOKE," Dr. D. Ewen Cameron and colleagues used these techniques to reduce many psychiatric patients to the "vegetable level." His mind-control experiments caused permanent brain damage, driving some Canadian and American subjects to madness and even suicide. Again, over half the subjects hallucinated; two became overtly psychotic; some suffered complete amnesia. Yet these studies continued over a decade, conducted by highly respected researchers.

Cameron's credentials were impressive: he was the founder and director of the Allan Memorial Institute, president of the Quebec Psychiatric Association, the Canadian Psychiatric Association, the American Psychiatric Association, and the World Psychiatric Association.

At the Allan Memorial Institute, Cameron used a technique he called psychic driving as a "therapeutic" method of "driving" patients with "verbal cues" to help "reorganize" their personalities. The vast majority of the human guinea pigs in these studies were "neurotic" women. In one of the studies disguised as "medical treatment," a woman who felt intensely rejected by her husband was forced to listen to repetitions of statements recorded in her own voice, including: "I can't count on my husband and my mother. . . . It makes me mad when I think of my past, when I was so lonely. . . . I am so lonely." Such psychic driving sessions went on for ten consecutive days of sixteen hours each. Cameron mildly dismissed his patients' "defenses against psychic driving" of bolting out of his office or trying to escape from the institute as "running away from the situation."

Suffering from depression after the birth of her daughter, Velma Orlikow (wife of a member of parliament in Canada) was hospitalized at the Allan Memorial Institute. Under the name of "medical treatment" she was isolated in an empty room, given LSD fourteen times, and forced to listen to painful taped messages for six hours a day. Today, she continues to suffer from chronic depression, takes drugs to sleep, and can no longer read—previously a favorite pastime. Others now suffer from amnesia, insomnia, the inability to read, and other mental and physical disabilities. And yet these experiments were supported by Canadian Department of National Health and Welfare and published in Canadian and American medical journals.

Other studies involved drugged sleep accompanied by massive electroshock treatments. After thirty to sixty shocks, the subjects were totally disoriented—they did not recognize anyone or know where they were, they were incontinent, and had difficulty performing simple motor skills. A ten-year follow-up study of these patients by other researchers showed that most of the patients suffered permanent memory loss. Electroconvulsive "therapy" had erased from six months to ten years of their experience. Although these other researchers recommended stopping intensive electroshock, it still continued.

While institutional review boards and a policy of informed consent have since been created to prevent abuse of experimental subjects, the research conducted by Cameron and Hoff was approved by the peer review panels that granted funding and recommended publication of the research. Two decades later, the psychiatric profession still refuses to acknowledge that an eminent leader in their field performed such unethical, destructive, and dehumanizing experiments on human beings in the name of science and medical treatment. At his death in 1967, Cameron was lauded by the *Canadian Psychiatric Association Journal* as a "diligent seeker after knowledge" who brought "deeper understanding of the importance and significance of the emotional life of man." It may seem ironic that a man with such underdeveloped feeling should devote his life to trying to understand the emotional life of man, but as suggested in the chapter on subjectivity, perhaps Cameron's research reflected an inner urge to develop his feeling function through the alchemical vessel of his work.

How is it that psychiatry, the field of medicine that is supposed to heal the psyche, comes to rape it instead? I do not mean to imply that psychological research is evil, but rather to point out that even the best of us can open ourselves to doing evil, with what we believe to be sincerely noble intentions. In reality, every form of consciousness produces its own shadow, its own kind of unconsciousness. It frequently creates the very evils it claims to despise. While Cameron sought to "understand the emotional life of man," he did so at great cost to the emotional life of patients entrusted to his care. He lacked the relatedness of the Feminine to connect him to his patients and stop him from pursuing experiments at their expense.

How do we rationalize such behavior? First of all we defend ourselves by denial, saying things like, "we're not doing permanent harm, they're exaggerating, the patients were already neurotic, it's for their own good, I wouldn't hurt anybody." Or we excuse it in the name of national security, the pursuit of knowledge, and so on, doing "whatever it takes" for the cause. For the greater good, we say that the end justifies the means. We can become possessed by goals such as national defense, feeding the hungry, helping developing countries, or curing disease. As we focus so intently on achieving the noble goal, we look away from the questionable steps it takes to get there. Such possession and narrowness of focus give us tremendous energy because no thought is wasted on ethical conflict. We often know when something is wrong, but splitting off our feeling into a separate compartment allows us to keep working, unencumbered by sorrow or guilt. Swept away by intellectual curiosity, and supremely confident in the power of thinking, scientists can step inadvertently into the most horrible evil without even noticing what they are doing.

As something that brings harm or suffering, evil in its simplest form is a fact of nature that we must overcome or escape—fight or flee. Few experience moral qualms about fighting off a shark about to devour us. An avalanche is so much bigger than we are that we have no choice but to flee. We do what we have to do in order to survive. But when different values come into conflict, we must make ethical choices. How do we balance the quest for knowledge with the desire to do no harm?

Even when we try to be moral and ethical, we all do bad things that we do not notice. When we do notice, we usually have an excuse: I had a headache; it was the other person's fault; I couldn't help myself; they asked for it; they deserved it; I was only doing my job; everyone else is doing it. Or we forget about the incident. But some people are more ethically sensitive than others. Guilt, bad dreams, or nightmares may haunt them over a transgression that does not faze a more ethically thick-skinned (or unconscious) person. As our feminine side comes more to light and we become increasingly conscious of the whole, of the interconnections between each other and between humanity and nature, we face more and more conflicts. We find we can no longer get by with such a shifty way of denying or excusing invasive or abusive behavior—even if the collective code condones it. The time comes when our personal ethics and integrity compel us to take a stand. Our feeling function and sense of relatedness connects us to the other and stops us. But we can get paralyzed or become martyrs if we try to be absolutely moral and ethical. We must each decide how far we can go. In struggling with conflicts of duty and values, we become more ethically sensitive and thus contribute to the evolution of consciousness.

Conventionally minded Westerners take their greatest pride in the achievements of modern science and technology. Yet it is where we are most sure about our perceptions, where we feel the most righteous, that evil is most likely to creep in. The power of the collective mind reinforces our most cherished ideas and opinions from within and without. It requires a tremendous effort of consciousness to question dogmas of orthodoxy and the behavior of authorities who have a strong sense of entitlement. It is also dangerous. But for the sake of consciousness, we must each take up the effort. As Abraham Maslow observed:

> It seems . . . that these "good," "nice" scientific words—prediction, control, rigor, certainty, exactness, preciseness, neatness, orderliness, lawfulness, quantification, proof, explanation, validation, reliability, organization, etc.—all are capable of being pathologized when pushed to the extreme. All of these same . . . goals are also found in the growth-motivated scientist. The difference is that they are not neuroticized. They are not compulsive, rigid, and uncontrollable. . . . They are not desperately needed, nor are they exclusively needed. It is possible for healthy scientists to enjoy not only the beauties of precision but also the pleasures of sloppiness, casualness, and ambiguity. . . . They are not afraid of hunches, intuitions, or improbable ideas.[6]

Life is an ambiguous business, and the business of science is no exception. None of us can fully own the swirling polarities within us and around us. The ethical task is to do our best and own the light and dark sides of our nature and not project upon others, despite the temptation. The greatest danger lies in our one-sided thinking; it is where we are most sure, most certain, that we are most vulnerable to doing evil. Rather than persist in the sanitized illusion that we can do no wrong as we pursue knowledge for the sake of knowledge, we can use our feeling function and the relatedness of the Feminine to help us evaluate each complex situation and research project in context—the hazards as well as the benefits.

SHINING LIGHT ON THE SHADOW

Scapegoating, projecting our shadow onto others, can justify the most extreme atrocities. In order to thwart the Devil's attempt to spread death and desolation among God's people, the Church tortured herbalists, midwives, and wise women suspected of witchcraft

277

until they confessed their crimes and denounced their "accomplices." Over 100,000 people were put to death in Europe between 1480 and 1700, 83 percent of whom were women.[7] Women were scapegoats for crop failures, storms, famine, illness, and male impotence. Another example comes from the 1950s, when Joseph McCarthy ruined many lives and careers in the name of saving America from communism. Likewise, during the Cold War, America projected its aggression onto the Soviets. Today, in a wave of "Japan-bashing," American business blames the Japanese for our economic decline.

Everything of evil tends to produce a chain reaction, whether it be revenge, paying back evil, or denouncing others under torture. When we have the courage to detach ourselves from the chain reaction and give up its power, we can interrupt it. For some, like Marti Crouch, this means leaving experimental science. Others try to create an oasis within their own personal sphere. For example, Julie Deans, a postdoc in immunology at the University of Washington, has decided that she must try to step aside from the evils that arise from corrosive competition and that result from some of the objectives and methods of science. She said:

> We can choose to withdraw from that dynamic and not compromise ourselves. I can only do it one relationship at a time. I can only influence those I'm in contact with. I get too overwhelmed if I try to change everyone or change the system. We can choose not to accept money from the military, not to work with animals. These are changes we can make now. To make any real changes might take drastic measures. I have heard that some women in Canada are setting up labs in their basements independent of public funding. They believe that the only way things will change in a meaningful way is to work with other women to create a completely different model of science.[8]

In collaborations, we can start a flow of generosity, giving our help and ideas. But if we find ourselves being exploited, we can turn off the tap and give nothing. Otherwise we are feeding the demon of that person or system, and contributing to the chain reaction. Integrity in such a situation may be more important than rational thinking, intelligence, or self-control.

In order to avoid eruptions of evil in science, we must withdraw our projections and take responsibility for our shadow. Until we can take back the burden of our own shadow, evil will continue to sneak out behind our backs, as it did with Ewen Cameron. As men liberate the Feminine in themselves, and withdraw their projections from women, they allow women to be themselves.

We can look to the alchemical process for guidance in the process of liberating the Feminine from the shadow of science. The first stage entails breaking down old definitions of science as a "masculine philosophy" and discarding beliefs that feeling and the relatedness of the Feminine are irrelevant to science. This involves acknowledging the contributions of women and openly exploring the value of the Feminine. As we focus the light of consciousness on each of its qualities, we can glean useful new perspectives.

Next, we experiment with new ways of combining thinking with feeling, aggression with receptivity, objectivity with subjectivity, multiplicity with hierarchy, personal with professional, competition with cooperation, sensation with intuition, analytical reductionism with relatedness. We can expect conflict and confusion to arise as the Masculine and Feminine begin to interact as equals. Deleterious unions of the opposites will abort. Finally, beneficial unions will give birth to an entirely new and wonderful science.

An example of this process can be found in a workshop about the future of chemistry and chemistry education. The faculty of a prestigious chemistry department asked psychiatrist and futurist Charles Johnston to facilitate their discussions. The faculty expressed the common frustration that the students were apathetic and not motivated for learning. One professor observed that students feel apathetic because they feel powerless.

At that point, another professor courageously withdrew the projection of powerlessness and admitted, "You know, I think we as faculty feel powerless." Recognizing a potentially creative moment in the pregnant pause that filled the room, Johnston asked, "How could that be? You are tenured faculty, men—not women— scientists, not in the humanities department. If anyone should feel empowered in the culture, it should be you. And yet you feel powerless."

They acknowledged that if the faculty did not feel inspired, excited, and empowered, neither would the students. Johnston

279

asked what it would take for the faculty members to feel empowered. After the rote answers of "more funding, more space, and more recognition," one professor said, "We need to talk to each other more. We're too isolated in our labs doing our own research. To feel a sense of power, meaning, and contribution in what we do, we need to confront the difficult and important questions about chemistry's contribution to the future. I think we've been frightened and reluctant to do that because every question in chemistry is now a moral question."

The discussion became animated as they realized that the emerging questions in chemistry were new kinds of questions—questions of value and moral questions about the potential of doing significant harm. Realizing that their training as chemists did not prepare them to deal with these questions, they discussed who they might bring into the conversation. They decided to invite professors from the fields of philosophy and literature to probe into questions of soul and purpose, and to create a story about the future of chemistry that would infuse their lives with meaning. All their apathy turned to excitement as they made plans for ongoing forums—which would include the students—to explore these new questions.[9]

In order to shine light onto the shadow of the institutions of science, we can consciously structure groups involved in evaluation processes to include individuals representing each of the Jungian psychological types. Feeling types could be welcomed and valued for their contributions as they seriously and constructively seek to raise ethical issues. Research teams, problem-solving task forces, brainstorming groups, grant-funding associations, peer review panels, policy-making councils, departmental staffs, ethics committees, informal scientific networks—all can benefit by opening themselves up to the disparate standpoints represented by the different functions. In doing so they can begin to withdraw their projections. Similarly, women, people of color, and people from different cultural backgrounds can be consciously included in evaluation processes.

Although we feel more comfortable amongst people similar to us, growth arises out of the lively debates that come from clashing worldviews. In order for such an exchange to be life-giving, it must be conducted in an atmosphere of respect—listening and receiving what the other has to say, rather than simply promoting and defending our own position. The strengths and weaknesses of each viewpoint will emerge as different assumptions and perspectives are examined, challenged, denied, projected, and finally integrated. It's an unsettling process as the ground quivers and shifts under the foundations of belief systems. We can each take the risk of standing on the threshold of change. Through such a process, both the individual and the institutions will change.

THE PATH TO RENEWAL

As opposed to the linear march of progress, the Great Round of the ancient mystery religions is a symbol of the opposites. It embraces the positive and negative, male and female, day and night, order and chaos, conscious and unconscious. Containing the totality, the rhythmic changes of the Great Round encompass the seasons of the year, the cycles of life and death, creation and destruction, the bearing of children and returning of corpses to the earth. Within the Great Round, the process of life does not consist of unchecked progress. Instead, it embodies the conflict between growth and decay, where growth is only one of its aspects.

The alchemical process involves suffering the conflict of the opposites until the creative solution is found, until something unexpected emerges that resolves the conflict on another level. This does not mean ignoring the problem and hoping a solution will easily appear, but rather completely developing all aspects of the conflict. Even as we proceed along the new path, however, we must remain aware of the risk of committing an error; we must give up our attitudes of certainty and control. As Marie-Louise von Franz expresses it:

> One is never sure, but from the Jungian standpoint it is better always to remain in an attitude of doubt towards one's own behavior, which means to do the best one can, but always to be ready to assume that one has made a mistake. This is a grown-up attitude which has given up clinging to infantile kindergarten rules.[10]

From the shadow, the Feminine has held a compensatory position to consciousness. The Feminine has always picked up what has been

281

neglected, overlooked, and disliked but which must still be examined and kept alive. In our lives as well as our research, the Feminine teaches us that the meaningful solution is always dependent upon context. It is individual. Through it we can begin the endless task of respecting life in all its uniqueness. Male or female, we can each have the courage to let the Feminine in us announce the truth in her own style and find out what she is trying to give us.

As modern-day alchemists, we can experiment with new ways of combining and recombining, synthesizing and separating, bringing together and integrating the masculine and feminine elements of ourselves. We can enlarge our consciousness beyond the roles of scientists as researchers, administrators, and professors, to become more active citizens, participants in the community, loving spouses, caring parents, and ordinary human beings as well as experts. Women do not need to adopt the stance of a woman indistinguishable from a man. With an appreciation of multiplicity without hierarchy, we can accept the differences symbolized as masculine and feminine as belonging to members of both sexes. As each person struggles to bring these elements together, we will find new and different ways of intertwining the personal and the professional.

Atmospheric scientist Kristina Katsaros feels she has had the best of both worlds as a wife and mother, and a scientist. Early in her career she worked part-time while she raised two children. Although she was in the background of science for many years, she felt grateful she could "do her thing" without the tension of feeling split between home and career. She could have a full life pursuing her intellectual interests, as well as "being a mama and enjoying it fully." She feels sorry for women who are pressured to perform in order to earn the right to exist in science, and for men who miss the chance to be close to their children. Now in her fifties, productive and well respected, she is full of enthusiasm for her work. But around her she sees middle-aged men who have been burned out by their struggle to make a name for themselves in science. Their enthusiasm is already waning, and they are tired. She observes, "To me it seems they don't have quite the joy that I do. They try to get out of things all the time. They don't want to be bothered as much. They only want bright students; they don't want to help anybody else; they're not patient anymore. It's kind of sad. Middle-aged men, old men."[11] In the generation of scientists now retiring, there was a different relationship between couples. While the men were in the lab establishing their careers, Katsaros sometimes saw their wives wither as they supported their husbands all the time, with no support of their own. In doing so, these women sacrificed their own potential development and sense of self. Women scientists rarely had that kind of buttressing (nor should such sacrifice of another be required to do science).

She contrasts the fast-paced, competitive American system to the socialist approach in Scandinavia, which has more concern for the human aspect. In Sweden both men and women have the right to take time off to be a parent, and daycare is readily available. In addition, Swedish scientists are more active in research concerning societal problems such as pollution, long-range transport, and carbon dioxide changes in climate. They can be more relaxed about their work and take off weekends, which benefits their health and their family. She finds the short-term American approach, where there are no guarantees for funding from year to year, to be wasteful, disrespectful, and unhealthy.

Physicist Eberhard Riedel echoes Katsaros's desire for more balance in science. He would like science as a discipline to expand, accepting the presence of women and the Feminine—without making a judgment of them as better or worse, but just there as part of the whole. His ideal image of a woman scientist comes from a book. He recalls:

> I read a book many years ago. I don't remember the title, but I remember this one scene of a scientist couple having a cocktail party of famous scientists. They had three kids who ran around, and now and then went back to their mother and she would hug them. It was wonderful that she felt so secure in her feminine, in her roles both as a mother and as a scientist, that she could do that. It's my dream of how a woman scientist could be—not pushing away something, not making a cut between science and personal, that the kids could come in and be part of it.[12]

And as men come to be more comfortable with nurturing, we can envisage the children also running to their father for a hug.

We now move into the second half of life as we face the potential destruction of life on Earth made possible by the products of science. With awareness of our mortality, we must negotiate the

midlife crisis of science. Rather than defend our ossifying old definitions of science, let us open ourselves to renewal. Let us welcome
the soul, the Feminine, to science. At midlife, consciousness tends
to continue in established attitudes and does not notice the inner
renewal taking place under the surface. Often renewal comes from
where we least expect it—from children, from simple people, or
from an officially despised part of the psyche such as the Feminine.

I do not believe in spurning technology or think that imposing
more bureaucratic layers will solve the problems of science. I see it as
a challenge to all individuals to open our minds to new possibilities,
to reflect deeply, to reexamine our values, to come to know ourselves, to develop our feeling and intuition to complement thinking
and sensation, to integrate the Feminine—to become more whole
people. Then each of us can begin to infuse science with heart and
humor. We can reach out to colleagues and build cooperative networks based on love, trust, and curiosity. In doing so, we become
living Philosopher's Stones. Everyone we touch with our lives will
see the value in this way of doing science. I believe in the power of the
small, the cumulative power of individuals leading conscious, ethical lives. As chaos science teaches us, once we reach a critical threshold, the institutions of science will reorganize themselves.

A poem by William Butler Yeats captures the joy of fully embracing life. In the alchemical dance of the Masculine and Feminine,
both the golden king and silver lady fully participate as individuals.
Together, they complement each other and create new harmonies.

> Of golden king and silver lady,
> Bellowing up and bellowing round,
> Till toes mastered a sweet measure,
> Mouth mastered a sweet sound,
> Prancing round and prancing up
> Until they pranced upon the top.
>
> That golden king and that wild lady
> Sang till stars began to fade,
> Hands gripped in hands, toes close together,
> Hair spread on the wind they made;
> That lady and that golden king
> Could like a brace of blackbirds sing.
>
> —*from* "Under the Round Tower"[13]

NOTES

PREFACE

1. T. McCormack, "Good Theory or Just Theory? Toward a Feminist Philosophy of Social Science," *Women's Studies International Quarterly* 4 (1981): pp. 1–12. McCormack comments that "as fashions in the historiography of science change, the qualities considered indispensable to excellence . . . change, but the subordinate, inferior position of women remains the same. . . . In an earlier period when the essential quality of the scientific mind was defined as analytic ability, women were thought to be unintellectual, deficient in reasoning ability. Warm and sensual, they were damned with faint praise for their allegedly 'natural' gift of intuitive insight, a desirable but clearly a lower level of skill for the heirs of Descartes. At present when the history of science is being rewritten in terms of creative, Kuhnian paradigmatic leaps, the brilliant scientific mind is described differently: a type of concentration that is loose, intuitive, a bit frivolous, if not wayward. Women who should be reaping the rewards of this revision are described as being overly cautious, too bound by experimental data, unwilling to speculate and, on the whole, too rational."

CHAPTER 1. VEILING THE FEMININE FACE OF SCIENCE

1. Aristotle, *On the Generation of Animals*, translated by A. L. Peck (Portsmouth, N.H.: Heinemann Educational Books, 1953), p. 11, §716a.
2. *The Politics of Aristotle*, translated by E. Barker (Oxford: Oxford University Press, 1946), p. 13, §1254b and p. 327, §1335b.
3. Aristotle, "On the Generation of Animals," in *The Works of Aristotle*, translated by Arthur Platt, from vol. 2 of *The Great Books of the Western World* (Chicago: William Benton, publisher for *Encyclopaedia Britannica*, 1952), p. 278, §737a.
4. Brian Easlea, *Witch Craft, Magic and the New Philosophy* (Atlantic Highlands, N.J.: Humanities Press, 1980), pp. 48–49.

Notes

5. Aristotle, "On the Generation of Animals," translated by Arthur Platt, p. 278, §737a.

6. *Webster's Ninth New Collegiate Dictionary* defines soul as "the animating principle, . . . the spiritual principle embodied in all human beings, . . . a man's moral and emotional nature, . . . the quality that arouses emotion and sentiment, . . . spiritual or moral force, . . . a strong positive feeling (as of intense sensitivity and emotional fervor)." Soul has also been projected onto blacks, as a characteristic of their culture. (Springfield, Mass.: Merriam-Webster, 1989), pp. 1126–1127.

7. Interview on January 11, 1990, with Ingrith J. Deyrup-Olsen, Professor in the Department of Zoology at the University of Washington in Seattle, Wash. She obtained her Ph.D. in 1944 and now studies general physiology and cell-membrane phenomena—in particular, the complexities of slug mucus.

8. Margaret Mead, *Sex and Temperament in Three Primitive Societies* (New York: William Morrow, 1935).

9. J. Needham, "History and Human Values: A Chinese Perspective for World Science and Technology," in H. Rose and S. Rose, eds., *Ideology of/in the Natural Sciences* (Cambridge, Mass.: Schenkman, 1976), pp. 255–256.

10. C. G. Jung, *The Archetypes and the Collective Unconscious*, vol. 9 in *The Collected Works of C. G. Jung*, translated by R. F. C. Hull (Princeton: Princeton University Press, 1957), p. 71, ¶147.

11. Sukie Colgrave, *Uniting Heaven and Earth: A Jungian and Taoist Exploration of the Masculine and Feminine in Human Consciousness* (Los Angeles: Jeremy P. Tarcher, 1979).

12. E. Haeckel's law of recapitulation stated that "ontogeny recapitulates phylogeny." In other words, embryonic development of an egg (ontogeny) or individual organism repeats the evolutionary development of the phyla (phylogeny). Because animals have evolved one phylum after another, Haeckel also proposed that "phylogeny causes ontogeny"—that their embryos still pass through this same succession of evolutionary stages. Biologists now realize this "law" is invalid and greatly oversimplifies evolution. More recent theories maintain that new evolution occurs by developmental diversion, in which a new path of embryonic or larval development branches away from some point along a preexisting developmental path.

13. Erich Neumann, *The Greater Mother: An Analysis of the Archetype*, translated by Ralph Manheim (Princeton: Princeton University Press, 1955), p. 43.

14. Julian Jaynes, *The Origin of Consciousness and the Breakdown of the Bicameral Mind* (Boston: Houghton Mifflin, 1976).

15. Erich Neumann, *The Origins and History of Consciousness*, translated by R. F. C. Hull (Princeton: Princeton University Press, 1954), pp. 140–144.

16. Polly Young-Eisendrath and Florence L. Wiedemann, *Female Authority: Empowering Women through Psychotherapy* (New York: Guilford Press, 1987).

17. Connie Zweig, ed., *To Be a Woman: The Birth of the Conscious Feminine* (Los Angeles: Jeremy P. Tarcher, 1990), p. 5.

18. Sigma Xi, the Scientific Research Society, *A New Agenda for Science* (New Haven, Conn.: Sigma Xi, 1986).

19. David F. Noble, "A World Without Women," *Technology Review* (May/ June 1992), pp. 53–60.

20. Joseph Glanvill, *The Vanity of Dogmatizing* (New York: Columbia University Press, reproduced for the Facsimile Text Society, 1931 [1661]), p. 118.

21. Ibid., p. 135.

22. Brian Easlea, *Witch Craft, Magic and the New Philosophy*, p. 214.

23. Londa Shiebinger, *The Mind Has No Sex* (Cambridge, Mass.: Harvard University Press, 1989), pp. 137–138, 276.

24. Francis Bacon, *Novum Organum*, vol. 4, p. 42, and *Of the Dignity and Advancement of Learning*, p. 373, in J. Spedding, R. L. Ellis and D. N. Heath, eds., *The Works of Francis Bacon* (London, 1858–61; reprinted Stuttgart: Friedrich Frommann Verlag, 1963).

25. B. Farrington, "Thoughts and Conclusions" in *The Philosophy of Francis Bacon* (Liverpool University Press, 1964), pp. 59, 62, 92, 93, 96.

26. See Carolyn Merchant's discussion in *The Death of Nature: Women, Ecology, and the Scientific Revolution* (New York: Harper & Row, 1980).

27. See Evelyn Fox Keller's analysis of the rhetoric of science in *Reflections on Gender and Science* (New Haven, Conn.: Yale University Press, 1985).

28. Ian Mitroff, *The Subjective Side of Science: A Philosophical Inquiry into the Psychology of the Apollo Moon Scientists* (Seaside, Calif.: Intersystems Publications, 1983), p. 210.

29. Murray Stein, *In MidLife* (Dallas: Spring Publications, 1983), p. 139.

CHAPTER 2. THE EMERGING VOICE OF THE FEMININE

1. I am grateful to Dr. Stephan Hoeller for his lectures and insights. For more information on a psychological interpretation of alchemy, refer to the following volumes in C. G. Jung's *Collected Works*: *Alchemical Studies* (vol. 13), *Psychology and Alchemy* (vol. 12), and *Mysterium Coniunctionis* (vol. 14) (Princeton University Press). Audio cassettes of Hoeller's lectures are available through BC Recordings, Box 2811, Los Angeles, CA 90078.

2. One of the earliest known alchemists was a woman, Cleopatra of Alexandria (not the queen). Although there were women alchemists in the medieval period, little is known about them. Men feared that inclusion of women would rob alchemy of respectability and leave practitioners of this transformational art vulnerable to accusations of witchcraft.

3. Thomas S. Kuhn, *The Structure of Scientific Revolutions* (Chicago: University of Chicago Press, enlarged second edition, 1970). The paradigm shift in thermodynamics is discussed on p. 67.

4. Carol Gilligan, *In a Different Voice: Psychological Theory and Women's Development* (Cambridge, Mass.: Harvard University Press, 1982).

5. Mary Field Belenky, Blythe McVicker Clinchy, Nancy Rule Goldberger, and Jill Mattuck Tarule, *Women's Ways of Knowing: The Development of Self, Voice, and Mind* (New York: Basic Books, 1986).

6. Carol Gilligan, "The Conquistador and the Dark Continent: Reflections on the Psychology of Love," *Daedalus* 113 (1984): p. 91.

7. William G. Perry, *Forms of Intellectual and Ethical Development in the College Years* (New York: Holt, Rinehart & Winston, 1970).

8. Belenky, *Women's Ways of Knowing*, p. 229.

9. For a review of the literature on gender and science, see Londa Shiebinger's "The History and Philosophy of Women in Science: A Review Essay," *Signs: Journal of Women in Culture and Society* 12, no. 2 (1987): pp. 305–332. Shiebinger identifies four conceptual approaches taken by various feminist scholars. The first approach seeks to recover the achievements of the neglected women scientists. The second examines women's limited access to the means of scientific production, the history of their participation in the institutions of science, and women's current status in the profession. The third approach analyzes how biological and medical sciences have defined the nature of women, telling us what is normal and natural. The fourth examines the distortions arising from the norms and methods of a science that excludes women.

10. Interview with Diane Horn on November 11, 1989, while she was a cell biologist in a cancer research division of Bristol-Myers Squibb, where she studied growth factors.

11. Margaret Rossiter, *Women Scientists in America: Struggles and Strategies to 1940* (Baltimore: Johns Hopkins University Press, 1984).

12. G. Kass-Simon and Patricia Farnes, *Women of Science: Righting the Record* (Bloomington, Ind.: Indiana University Press, 1990).

13. Kass-Simon, *Women of Science*, p. xiii.

14. William Booth, "Oh, I Thought You Were a Man," *Science* 243 (27 January 1989): p. 475. Physicist Lise Meitner shared the Enrico Fermi Award in 1966 with chemists Otto Hahn and Fritz Strassmann for research that led to the discovery of uranium fission.

15. L. M. Jones, "Intellectual Contributions of Women to Physics," in Kass-Simon, *Women of Science*, pp. 200–203.

16. Vera Kistiakowsky, "Women in Physics: Unnecessary, Injurious and Out of Place?" *Physics Today* 33, no. 2 (February 1980): p. 32.

17. Martin Goldman and Marian Gordon Goldman, "Will She Make It?," *Working Woman* 9 (January 1984): p. 104.

18. Dennis Overbye, "Einstein in Love," *Time* (30 April 1990): p. 108.

19. Carole Bodger, "Salary Survey: Who Does What and for How Much?," *Working Woman* (January 1985): p. 72.

20. America's most prestigious scientific society elected its first two women in 1925 and 1931. The oldest permanent scientific society, the Royal Society of London, excluded women until 1945. Even though discrimination became illegal in the U.S. in 1964, women still live in a "chilly climate" within the system.

21. In an essay opposing the appointment of a woman mathematician to a professorship at the University of Stockholm at the end of the nineteenth century, the author tries to prove "as decidedly as that two and two make four, what a monstrosity is a woman who is a professor of mathematics, and how unnecessary, injurious and out of place she is." H. J. Mozans, *Women in Science* (Notre Dame, Ind.: University of Notre Dame Press, 1991), pp. 162–163.

22. National Science Foundation's study of employed doctoral scientists in *Professional Women and Minorities: A Manpower Data Resource Service*, compiled by Betty M. Vetter and Eleanor L. Babco (Commission of Professionals in Science and Technology, December 1987), p. 95. In comparison, only 3.6 percent of physicists/astronomers, 8.7 percent of chemists, and 2.2 percent of engineers are women.

23. Ruth Hubbard, *The Politics of Women's Biology* (New Brunswick, N.J.: Rutgers University Press, 1990), p. 137.

24. Marian Lowe, "Dialectics of Biology and Culture," in Lowe and Hubbard, eds., *Women's Nature: Rationalizations of Inequality* (Elmsford, N.Y.: Pergamon Press, 1983), pp. 39–62. Lowe references work by Jack H. Wilmore, "Inferiority of Female Athletes, Myth or Reality," *Journal of Sports Medicine* 3 (1975): pp. 1–6.

25. Eleanor Maccoby and Carol Nagy Jacklin, *The Psychology of Sex Differences* (Stanford, Calif.: Stanford University Press, 1974).

26. Anne Fausto-Sterling, *Myths of Gender* (New York: Basic Books, 1985), pp. 53–59.

27. Jeanne M. Stellmann and Mary Sue Henifin, "No Fertile Women Need Apply: Employment Discrimination and Reproductive Hazards in the Workplace," in Ruth Hubbard, Mary Sue Henifin, and Barbara Fried, eds., *Biological Woman—The Convenient Myth* (Rochester, Vt.: Schenkman Books, 1982), pp. 117–146.

28. Geoffrey Sea, "Radiation and Response: Dose, Disease and the Development of Health Physics," presented at the History of Science Society conference, October 25–28, 1990.

29. R. Dawkins, *The Selfish Gene* (New York: Oxford University Press, 1976), p. 176.

30. An expert in insect behaviors, E. O. Wilson sought to establish sociobiology "as the systematic study of the biological basis of all social behavior." Those in the Wilsonian school of human sociobiology believe that all human behaviors, social relationships, and organizations are genetically evolved adaptations.

31. E. O. Wilson, *On Human Nature* (Cambridge, Mass.: Harvard University Press, 1978), p. 125.

32. Jane van Lawick Goodall, *In the Shadow of Man* (Boston: Houghton Mifflin, 1971), pp. 79–88.

33. R. Easton, ed., "The World's Cats II" (Seattle, Wash.: Feline Research Group, Woodland Park Zoo, 1976).

34. Sarah Blaffer Hrdy, "Empathy, Polyandry, and the Myth of the Coy Female," in Ruth Bleier, ed., *Feminist Approaches to Science* (New York: Pergamon Press, 1988), pp. 139–140.

35. G. Schatten and H. Schatten, "The Energetic Egg," *The Sciences* 23, no. 5 (1983): pp. 28–34.

36. The Biology and Gender Study Group, "The Importance of Feminist Critique for Contemporary Cell Biology," in Nancy Tuana, ed., *Feminism and Science* (Bloomington, Ind.: Indiana University Press, 1989), p. 177.

CHAPTER 3. FEELING: RESEARCH MOTIVATED BY LOVE

1. Richard H. Lampkin, Jr., "Scientific Attitudes," *Science Education* 22, no. 7 (December 1938): p. 356.

2. Robert L. Ebel, "What Is the Scientific Attitude?" *Science Education* 22, no. 2 (February 1938): p. 78.

3. Mitroff, *The Subjective Side of Science*, pp. 114–116.

4. Ibid., p. 130.

5. Interview on Janaury 22, 1991, with Eberhard K. Riedel, Professor of Physics at the University of Washington in Seattle, Washington. Riedel received his Ph.D. in 1966 and studies theoretical condensed-matter physics.

6. In "Scientific Attitudes," p. 354, Lampkin describes the steps of the scientific method as "(1) experiencing sensation, (2) classification of sense data, and (3) the elimination of any classifications whose implications are incompatible with the sense data."

7. The MBTI is commercially available through Consulting Psychologists Press, Inc., 3803 E. Bayshore Road, Palo Alto, CA 94303.

8. Deriving from the legacy of Greek philosophy, Western science is founded on the "rational conception of the cosmos as an orderly whole working by laws discoverable in thought." See *The Edge of Objectivity* by Charles Coulston Gillispie (Princeton: Princeton University Press, 1960), p. 9.

9. "The Nature of Science," chapter 1 from the 1989 AAAS report *Science for All Americans*, published in *Bulletin of Science, Technology, and Society* 10, no. 2 (1990): p. 95.

10. Interview on March 14, 1991, with Peggy Johnson (a pseudonym chosen to protect the confidentiality of the peer review process), Senior Staff Scientist at a biotechnology company. She received her Ph.D. in biochemistry in 1980.

11. Carol Cohn, "Sex and Death in the Rational World of Defense Intellectuals," *Signs: Journal of Women in Culture and Society* 12, no. 4 (1987): pp. 687–718.

12. Interview with Jungian analyst Anne de Vore on May 30, 1991. She earned a B.A. in philosophy and literature, an M.A. in counseling, and a Ph.D. in educational psychology; she is also a Diplomate of the Inter-Regional Society of Jungian Analysts, and a nationally certified psychoanalyst.

13. Interview on January 22, 1991, with Eberhard K. Riedel.

14. Robert L. Sinsheimer, "The Presumptions of Science," *Daedalus* 107, no. 2 (Spring 1978—this issue is entitled "Limits of Scientific Inquiry"): pp. 23–35. *Daedalus* is the house organ of the American Academy of Arts and Sciences. Quoted by Ruth Hubbard in *Politics of Women's Biology* (New Brunswick, N.J.: Rutgers University Press, 1990), p. 11.

15. Robert S. Morison, "Introduction," *Daedalus* 107, no. 2 (Spring 1978—the "Limits of Scientific Inquiry" issue): pp. vii–xvi. Quoted by Ruth Hubbard in *Politics of Women's Biology*, p. 10.

16. Interview on March 8, 1991, with Marsha Landolt, Professor of Fisheries in the College of Ocean and Fishery Sciences at the University of Washington in Seattle. She received her Ph.D. in 1976 and studies fish and shellfish diseases. She is currently Director of the School of Fisheries.

17. Ibid.

18. Ibid.

19. Interview on March 14, 1991, with Peggy Johnson.

20. Ibid.

21. Ethlie Ann Vare and Gregg Ptacek, *Mothers of Invention: From the Bra to the Bomb, Forgotten Women and Their Unforgettable Ideas* (New York: William Morrow & Co., 1988), pp. 150–152.

Notes

22. Sue V. Rosser, *Female-Friendly Science: Applying Women's Studies Methods and Theories to Attract Students* (New York: Pergamon Press, 1990), p. 42.
23. R. Cowen, "President's Budget: Rosy Outlook for R&D," *Science News* 139, no. 6 (9 February 1991): pp. 87, 94.
24. "Where love reigns, there is no will to power; and where the will to power is paramount, love is lacking." C. G. Jung, *Two Essays on Analytical Psychology*, vol. 7 in *The Collected Works of C. G. Jung*, translated by R. F. C. Hull (Princeton: Princeton University Press, 1959), p. 53, ¶ 78.
25. Belenky, *Women's Ways of Knowing*, p. 141.
26. Keller, *Reflections on Gender and Science*, pp. 52–53.
27. Quoted by Carolyn Merchant in *The Death of Nature* (New York: Harper & Row, 1980), p. 104, from Giovanni Battista della Porta, *Magia Naturalis* (Naples, 1558). English translation, Derek J. Price, ed., *Natural Magic* (facsimile ed., New York: Basic Books, 1957; first published 1658), p. 14.
28. Charles Singer, ed., *Studies in the History and Method of Science* (Oxford: Clarendon Press, 1921), p. 188.
29. Interview on March 11, 1991, with Becca Dickstein, Assistant Professor in the Department of Bioscience and Biotechnology at Drexel University. She received her Ph.D. in 1985 and studies symbiosis between legumes and nitrogen-fixing bacteria.
30. Interview with Sigrid Myrdal on November 11, 1989. After holding two postdoctoral positions, she obtained a position in a cancer research division of Bristol-Myers Squibb, where she studies growth factors.
31. Interview on January 11, 1990, with Ingrith J. Deyrup-Olsen.
32. Evelyn Fox Keller, *A Feeling for the Organism: The Life and Work of Barbara McClintock* (San Francisco: W. H. Freeman, 1983).
33. Keller, *Reflections on Gender and Science*, p. 164.
34. Ibid., p. 165.
35. Evelyn Fox Keller, *A Feeling for the Organism*, pp. 205–206.
36. Brian Easlea, *Witch Craft, Magic and the New Philosophy*, p. 214.
37. Ibid.
38. Shirley Briggs, "Rachel Carson: Her Vision and Her Legacy," in Gino Marco, Robert Hollingworth, and William Durham, eds., *Silent Spring Revisited* (Washington, D.C.: American Chemical Society, 1987), p. 4.
39. Quoted from Kawai Masao's book *Life of the Japanese Monkeys* in Sy Montgomery, *Walking with the Great Apes: Jane Goodall, Dian Fossey, Biruté Galdikas* (Boston: Houghton Mifflin, 1991), p. 275.
40. Kevin W. Kelley, ed., *The Home Planet* (Reading, Mass.: Addison-Wesley Publishing Company, 1988), p. 60.

41. Interview on February 4, 1991, with Kristina Katsaros, Professor of Atmospheric Sciences at the University of Washington in Seattle. She obtained her Ph.D. in 1969 and studies air-sea interactions.

42. Interview on February 13, 1991, with Cynthia Haggerty (a pseudonym). She received her Ph.D. in 1971 and worked for ten years at a federal marine biology laboratory with a joint appointment as adjunct professor in the Department of Zoology at the local university. In 1981 she was elected to be an AAAS Fellow.

CHAPTER 4. RECEPTIVITY: LISTENING TO NATURE

1. Brian Easlea, *Witch Craft, Magic and the New Philosophy*, p. 214. Robert Boyle was one of the founding members of the Royal Society of London. He described the relation concerning the compression and expansion of a gas at room temperature that is now known as Boyle's law.

2. Thomas Kuhn, *The Structure of Scientific Revolutions*, p. 116.

3. Interview on February 1, 1991, with Paula Szkody, Research Professor in the Department of Astronomy at the University of Washington in Seattle. She received her Ph.D. in 1975 and her research interests include cataclysmic variables, photometry, and spectroscopy.

4. Interview on January 22, 1991, with Eberhard K. Riedel.

5. Interview on Janaury 17, 1990, with Aimee Bakken, Associate Professor in the Department of Zoology at the University of Washington in Seattle. She received her Ph.D. in 1970 and specializes in developmental and cell biology, developmental genetics, and chromosome structure and function in oogenesis and embryogenesis.

6. Interview on February 13, 1991, with Cynthia Haggerty.

7. Interview with Sigrid Myrdal on November 11, 1989.

8. James E. Lovelock, "Small Science" in John Brockman, ed., *Doing Science: The Reality Club* (New York: Prentice-Hall, 1988), p. 186.

9. Interview on January 11, 1990, with Ingrith J. Deyrup-Olsen.

10. Interview with Sigrid Myrdal on November 11, 1989.

11. Interview on November 11, 1989, with Sylvia Pollack, Research Professor in the Department of Biological Structure at the University of Washington in Seattle. She received her Ph.D. in 1967 and studies cellular immunology. In September 1990 she began a Masters degree program in psychology.

12. Karl Popper of the University of London propounds the doctrine of falsibility, where a scientific theory can never be proved true, it can only be refuted, and once refuted it is abandoned.

13. Interview on January 14, 1990, with Davida Y. Teller, Professor in the Department of Psychology, with joint appointments in the Departments of Physiology/Biophysics and the Women's Studies Program at the Univer-

sity of Washington in Seattle. She received her Ph.D. in 1965 and studies vision, the development of vision, and the philosophy of visual science.

14. Paul David Hamilton's phrase.

15. Thomas Kuhn, *The Structure of Scientific Revolutions*, p. 59. Kuhn gives numerous examples of how the predominant worldview precluded scientists from seeing nature.

16. "A Book for Burning?" *Nature* 293 (24 September 1981): pp. 245–246.

17. Brian Josephson, letter to the editor in *Nature* 293 (15 October 1981): p. 594.

18. William Broad and Nicholas Wade, *Betrayers of the Truth* (New York: Simon & Schuster, 1982), pp. 141–142.

19. Beginning in 1773, Laplace devoted his life to developing a completely mechanical interpretation of the solar system using Newton's laws of gravity. He also laid the mathematical foundations for the scientific study of heat, electricity, and magnetism.

20. Benoit B. Mandelbrot, *The Fractal Geometry of Nature* (New York: W. H. Freeman, 1977), p. 3.

21. Mandelbrot, *The Fractal Geometry of Nature*, pp. 20, 193, 422.

22. C. G. Jung, *Psychological Types*, vol. 6 in *The Collected Works of C. G. Jung*, a revision by R. F. C. Hull of the translation by H. G. Baynes (Princeton: Princeton University Press, 1971), pp. 426–427.

23. For a comparison of the Babylonian story of Tiamat with the Taoist story of Hun-tun, see Eugene Eoyang's "Chaos Misread: Or, There's a Wonton in My Soup," *Comparative Literature Studies* 26, no. 3 (1989): pp. 271–284.

24. N. Katherine Hayles, *Chaos and Disorder: Complex Dynamics in Literature and Science* (Chicago: University of Chicago Press, 1991), p. 6. Hayles quotes Ilya Prigogine and Isabelle Stengers, *Order Out of Chaos: Man's New Dialogue with Nature* (New York: Bantam, 1984).

25. Discussed by N. Katherine Hayles, *Chaos and Disorder*, p. 18.

26. H. J. Mozans, *Women in Science* (Notre Dame, Ind.: University of Notre Dame Press, 1991), pp. 162–163.

27. James Gleick, *Chaos: Making a New Science* (New York: Viking, 1987), p. 298.

CHAPTER 5. SUBJECTIVITY: DISCOVERING OUR SELVES THROUGH THE EXPERIMENT

1. Deepak Chopra, M.D., "The Quantum Mechanical Body," a lecture given at the American Holistic Medical Association Conference on March 30, 1990, in Seattle. Audio tapes of this lecture (HM30) are available through Sounds True Catalog, 1825 Pearl Street, Boulder, CO 80302. An endocrinologist and President of the American Association for

Ayurvedic Medicine, Chopra was formerly chief of staff of the New England Memorial Hospital.

2. Colin Blakemore and Grahame F. Cooper, "Development of the Brain Depends on the Visual Environment," *Nature* 228 (31 October 1970): pp. 477–478. Also see Helmut V. B. Hirsch and D. N. Spinelli, "Visual Experience Modifies Distribution of Horizontally and Vertically Oriented Receptive Fields in Cats," *Science* 168 (15 May 1970): pp. 869–871. I wish to thank Davida Teller for bringing these papers to my attention.

3. M. Polanyi, *The Logic of Liberty: Reflections and Rejoinders* (London: Routledge & Kegan Paul, 1951), p. 19.

4. M. von Senden, *Space and Sight: The Perception of Space and Shape in the Congenitally Blind Before and After Operation*, translated by Peter Heath (Glencoe, Ill.: Free Press, 1960), pp. 141, 144, 170.

5. Solomon E. Asch, "Opinions and Social Pressure," *Scientific American* 193, no. 5 (November 1955): pp. 31–35.

6. Interview on January 11, 1990, with Ingrith J. Deyrup-Olsen.

7. Israel Scheffler, *Science and Subjectivity* (Indianapolis: Bobbs-Merrill Company, 1967), p. 8.

8. Interview on February 4, 1991, with Kristina Katsaros.

9. Sharon Traweek, *Beamtimes and Lifetimes: The World of High Energy Physicists* (Cambridge, Mass.: Harvard University Press, 1988), p. 91.

10. Interview on January 2, 1990, with Patricia Thomas, Associate Professor at the Oklahoma Medical Research Foundation in Oklahoma City.

11. Daryl E. Chubin and Edward J. Hackett, *Peerless Science: Peer Review and U.S. Science Policy* (Albany: State University of New York Press, 1990), pp. 69–70. This book provides an excellent in-depth analysis and critique of the peer review process.

12. Of all doctoral scientists and engineers, 31.4 percent were employed by business or industry in 1985; 52.9 percent by educational institutions (with research generally supported by government grants or industrial contracts); 9.1 percent by state and federal governments; 2.8 percent by hospitals and clinics; and 3.4 percent by nonprofit organizations. See table 4-14 compiled by the National Research Council, in Vetter and Babco, *Professional Women and Minorities: A Manpower Data Resource Service*, p. 100.

13. From a panel discussion, "Whose Science Is It, Anyway?," sponsored by Puget Sound Science Writers Association on November 13, 1991, at the University of Washington in Seattle.

14. Belenky, *Women's Ways of Knowing*, p. 141.

15. Barbara Du Bois, "Passionate Scholarship: Notes on Values, Knowing, and Method in Feminist Social Science," in G. Bowles and R. Duelli-

Klein, eds., *Theories of Women's Studies* (Boston: Routledge & Kegan Paul, 1983), pp. 105–116.

16. Interview on December 1, 1991, with Sylvia Pollack.

17. Belenky, *Women's Ways of Knowing*, p. 146.

18. Marilyn Ferguson is the editor of *Brain/Mind Bulletin* and *Common Sense*.

19. Belenky, *Women's Ways of Knowing*, p. 141.

20. I wish to thank Lorraine Daston for her paper "The Objectivity of Interchangeable Observers, 1830–1900," presented at the 1990 History of Science Society Meeting in Seattle. Peter Dear's paper, "From Truth to Disinterestedness in the Seventeenth Century," and Theodore Porter's paper, "Quantification and the Accounting Ideal in Science," both from this meeting as well, also enriched my understanding of the history of objectivity in science.

21. Both Darwin and Wallace are quoted in Ruth Hubbard's essay, "Have Only Men Evolved?" in Sandra Harding and Merrill B. Hintikka, eds., *Discovering Reality: Feminist Perspectives on Epistemology, Metaphysics, Methodology, and Philosophy of Science* (Boston: D. Reidel Publishing Company, 1983), p. 51.

22. Theodore Porter, "Quantification and the Accounting Ideal in Science."

23. Founded in 1973, the Institute of Noetic Sciences is a nonprofit public research foundation, an educational institution, and a membership organization. The word *noetic* is derived from the Greek word *nous*, meaning mind, intelligence, or ways of knowing. For more information write to 475 Gate Five Road, Suite 300, Sausalito, CA 94964.

24. Scheffler, *Science and Subjectivity*, pp. v–vi.

25. "On Becoming a Scientist," published by the National Academy of Science's Committee on the Conduct of Science (Washington, D.C.: National Academy Press, 1989), p. 1.

26. Mitroff, *The Subjective Side of Science*, p. 65.

27. Ibid., p. 66.

28. Brian Martin, *The Bias of Science* (Canberra: Society for Social Responsibility in Science, 1979).

29. John Gribbin, *In Search of Schrödinger's Cat: Quantum Physics and Reality* (Toronto: Bantam Books, 1984), pp. 123–152.

30. Werner Heisenberg, *Physics and Philosophy* (New York: Harper & Row, 1958), p. 145.

31. The best description of the Aspect experiment for the lay reader can be found in Gribbin, *In Search of Schrödinger's Cat: Quantum Physics and Reality*, pp. 227–231.

32. Max Born, *Physics in My Generation* (London and New York: Pergamon Press, 1956), p. 48.

33. Donald Michael discusses the role of uncertainty in leadership in the videotape "Governance, Uncertainty, and Compassion" from the series *Thinking Allowed* (Sausalito, Calif.: Institute of Noetic Sciences, 1988).

34. Mandelbrot, *The Fractal Geometry of Nature*, p. 27.

35. Jung, *Psychological Types*, p. 457, ¶783.

36. Ibid., p. 9, ¶9.

37. Conservation with Anne de Vore, October 24, 1991.

CHAPTER 6. MULTIPLICITY: WEBS OF INTERACTION

1. J. I. Rodale, ed., *The Synonym Finder* (Emmaus, Penn.: Rodale Press, 1978), p. 506.

2. Deborah Tannen, *You Just Don't Understand: Women and Men in Conversation* (New York: William Morrow, 1990).

3. A comment made by a graduate student in forestry at a Women, Science, and Technology Reading and Discussion Group in the fall of 1990 at the University of Washington in Seattle.

4. Hilary Roberts, "A Qualified Failure," *New Scientist* 9 (June 1983): p. 722.

5. For the complete story, see Broad, *Betrayers of the Truth*, pp. 143–149.

6. Traweek, *Beamtimes and Lifetimes*, p. 91.

7. Ibid., p. 90.

8. Ibid., p. 88.

9. Ibid., pp. 25, 27–28.

10. Keller, *Reflections on Gender and Science*, p. 132.

11. J. T. Johnson, "Fuzzy Logic," *Popular Science* (July 1990): pp. 87–89. Lofti A. Zadeh at the University of California at Berkeley developed fuzzy-logic theory with the expectation that it would be applied to systems in which human judgment and emotions played a role—areas that lacked quantitative techniques, such as law and psychology.

12. C. G. Jung, *On the Nature of the Psyche*, from vol. 8 in *The Collected Works of C. G. Jung*, translated by R. F. C. Hull (Princeton: Princeton University Press, 1971), p. 142, ¶440.

13. Stephen Hawking, "Is the End in Sight for Theoretical Physics?," in the appendix to *Stephen Hawking's Universe: An Introduction to the Most Remarkable Scientist of Our Time* by John Boslough (New York: Avon Books, 1985).

14. Richard Wolkomir, "Quark City," *Omni* 6, no. 5 (February 1984): p. 41.

15. Grégoire Nicolis and Ilya Prigogine, *Exploring Complexity: An Introduc-*

tion (New York: W. H. Freeman, 1989), p. 13. Prigogine received the 1977 Nobel Prize for chemistry for his contributions to nonequilibrium thermodynamics, particularly the theory of dissipative structures.

16. From the 1990 workshop, "Artificial Life II," that took place in Santa Fe, New Mexico, described in "Spontaneous Order, Evolution, and Life," *Science* 247 (30 March 1990): pp. 1543–1544.

17. Carl Jung, in the foreword to Erich Neumann's *Origins and History of Consciousness* (Princeton: Princeton University Press, 1954), p. xiv.

18. Interview on November 11, 1989, with Sylvia Pollack.

19. Traweek, *Beamtimes and Lifetimes*, pp. 76–77.

20. Ibid., pp. 91–92.

21. Marion Namenwirth, "Science Seen Through a Feminist Prism" in Bleier, *Feminist Approaches to Science*, p. 23.

22. Harry F. Harlow, *Learning to Love* (New York: Jason Aronson, 1974), pp. 159–160.

23. Traweek, *Beamtimes and Lifetimes*, pp. 147–148.

24. Julius A. Roth, "Hired Hand Research," *The American Sociologist* (August 1966): pp. 190–196.

25. Broad, *Betrayers of the Truth*, p. 150.

26. C. G. Jung, *Psychology and Alchemy*, vol. 12 in *The Collected Works of C. G. Jung*, translated by R. F. C. Hull (Princeton: Princeton University Press, 1953), p. 28, ¶34.

27. Tannen, *You Just Don't Understand*, pp. 153–159.

28. From the program "A. Einstein: How I See the World," a 1991 production from Videfilm Producers International, Ltd., and Lumen Productions in association with public television station WNET.

29. Interview on February 13, 1991, with Cynthia Haggerty.

30. Ibid.

31. Ibid.

32. Interview on February 4, 1991, with Kristina Katsaros.

33. Shiebinger, *The Mind Has No Sex*, pp. 30–32.

34. Jane van Lawick Goodall, *In the Shadow of Man*, p. 6.

35. David Ehrenfeld, "The Next Environmental Crisis," *Conservation Biology* 3, no. 1 (March 1989): pp. 1–3.

36. W. A. Cooper and E. N. Walker, *Getting the Measure of the Stars* (Philadelphia: Adam Hilger, 1989).

37. Earthwatch, 680 Mt. Auburn Street, P.O. Box 403, Watertown, MA 02272.

38. World Wildlife Fund, 1250 Twenty-Fourth Street NW, P.O. Box 96220, Washington, DC 20037.

39. From discussions at the weekly Women, Science, and Technology Reading and Discussion Group held at the University of Washington in Seattle during the autumn of 1990.
40. Ibid. Katy Gray is a pseudonym.
41. Howard Youth, "Iguana Farms, Antelope Ranches," *World Watch* 4, no. 1 (January/February 1991): pp. 37–39.

CHAPTER 7: NURTURING: A LONG-TERM APPROACH

1. Traweek, *Beamtimes and Lifetimes*, p. 33.
2. Interview on January 17, 1990, with Aimee Bakken.
3. Nancy Griffith-Marriott, "Bodymind: An Interview with Candace Pert on Science, Feminism, Spirituality, and AIDS," *Woman of Power* 11 (Fall 1988): p. 25.
4. N. H. Bass, A. Hess, A. Pope, and C. Thalheimer, "Quantitative Cytoarchitectonic Distribution of Neurons, Glia and DNA in Rat Cerebral Cortex," *Journal of Comparative Neurology* 143: pp. 481–490.
5. Marian C. Diamond, Arnold B. Scheibel, Greer M. Murphy, Jr., and Thomas Harvey, "On the Brain of a Scientist: Albert Einstein," *Experimental Neurology* 88 (1985): pp. 198–204. Also discussed in "Fires of the Mind," a program in the series *The Infinite Voyage*, PTV Publications, P.O. Box 701, Kent, OH 44240.
6. Interview on January 11, 1990, with Ingrith J. Deyrup-Olsen.
7. Interview on December 4, 1990, with Aimee Bakken.
8. Ibid.
9. Interview on January 17, 1990, with Aimee Bakken.
10. Lewis Wolpert and Alison Richards, *A Passion for Science* (Oxford: Oxford University Press, 1988), p. 5.
11. Tannen, *You Just Don't Understand*, pp. 75–77.
12. Robert Gallo, *Virus Hunting: AIDS, Cancer, and the Human Retrovirus: A Story of Scientific Discovery* (New York: Basic Books, 1991), p. 165.
13. Interview on February 13, 1991, with Cynthia Haggerty.
14. From a conversation with Aimee Bakken.
15. Interview with Diane Horn on November 11, 1989.
16. Personal communication with Becca Dickstein, May 5, 1992.
17. Interview on February 4, 1991, with Kristina Katsaros.
18. Ibid.
19. Sally Macdonald, "Publish or Perish Throttles Teaching Role, Author Says," *The Seattle Times* (8 April 1990): p. B3.
20. David P. Hamilton, "Publishing by—and for?—the Numbers," *Science* 250 (7 December 1990): pp. 1331–1332.

Notes

21. Jennie Dusheck, "Female Primatologists Confer—Without Men," *Science* 249 (28 September 1990): pp. 1494–1495.

22. Ibid.

23. Alison Galloway, "All Women Conference: Did It Discriminate?" *Science* 249 (7 December 1990): p. 1319. Also personal communication with Alison Galloway. In a joint letter to *Science*, the organizers of the conference point out that the participants represented a broad spectrum of fields, including human physiology, reproductive endocrinology, ethnology, psychology, paleontology, functional morphology, behavioral ecology—not just primatology as the title to Dusheck's article implies.

24. Joel N. Shrukin, "Sexism and Hypocrisy," *Science* 249 (16 November 1990): p. 887.

25. Interview on January 17, 1990, with Aimee Bakken.

26. Ibid.

27. Montgomery, *Walking with the Great Apes: Jane Goodall, Dian Fossey, Biruté Galdikas*, p. 81.

CHAPTER 8. COOPERATION: WORKING IN HARMONY

1. Charles Darwin, "The Linnean Society Papers," in Philip Appleman, ed., *Darwin: A Norton Critical Edition* (New York: Norton, 1970), p. 83.

2. Alfred R. Wallace, "The Linnean Society Papers," p. 92.

3. Charles Darwin, *The Origin of Species*, 6th ed. (London, 1872; reprinted New York: Mentor, 1958), p. 74.

4. Robert Augros and George Stanciu, *The New Biology: Discovering the Wisdom in Nature* (Boston: Shambhala Publications, New Science Library, 1988).

5. Paul Colinvaux, *Why Big Fierce Animals Are Rare: An Ecologist's Perspective* (Princeton: Princeton University Press, 1978), p. 146.

6. Charles Fowler, "Comparative Population Dynamics in Large Animals," in Fowler and Smith, eds., *Dynamics of Large Mammal Populations* (New York: Wiley, 1981): pp. 444–445.

7. Norman Owen-Smith, "Territoriality in the White Rhinoceros (*Ceratotherium simium*) Burchell," *Nature* 231 (4 June 1971): p. 294.

8. "Corn's Weedy Companion," *The Seattle Times* (3 September 1990): p. D2.

9. Frits W. Went, "The Plants (New York: Time-Life Books, 1963), p. 168.

10. Frits W. Went, "The Ecology of Desert Plants," *Scientific American* 192 (April 1955): p. 74.

11. Robert Axelrod and William D. Hamilton, "The Evolution of Cooperation," *Science* 211 (27 March 1981): p. 1391.

12. Robert M. May, "A Test of Ideas about Mutualism," *Nature* 307 (February 1984): p. 410.
13. David Kirk, ed., *Biology Today* (New York: Random House, 1975), pp. 658–659.
14. Lynn Margulis, *Symbiosis in Cell Evolution* (San Francisco: W. H. Freeman, 1981), p. 164.
15. Charles Mann, "Lynn Margulis: Science's Unruly Earth Mother," *Science* 252 (19 April 1991): p. 379.
16. Lynn Margulis, "Words as Battle Cries—Symbiogenesis and the New Field of Endocytobiology," *BioScience* 40 (October 1990): pp. 675–676.
17. Ibid., p. 675.
18. Interview on March 11, 1991, with Becca Dickstein.
19. C. Ezzell, "Helping Cancers Mature So They Might Die," *Science News* 139 (1 June 1991): p. 341.
20. Sigma Xi, *A New Agenda for Science*, p. 43.
21. Daniel E. Koshland, Jr., editor of *Science*, "Waste Not, Want Some," *Science* 252 (26 April 1991): p. 485.
22. Leslie Roberts, "The Rush to Publish," *Science* 251 (January 18, 1991): pp. 260–263.
23. William J. Broad, "Imbroglio at Yale (1): Emergence of a Fraud," *Science* 210 (1980): pp. 38–41.
24. James Watson, "The Dissemination of Unpublished Information," in Saul Bellow, ed., *The Frontiers of Knowledge* (Garden City, N.Y.: Doubleday and Co., 1975), p. 161.
25. Sigma Xi, *A New Agenda for Science*, p. 43.
26. Quoted in Traweek, *Beamtimes and Lifetimes*, pp. 89–90.
27. Suzanne Gordon, *Prisoners of Men's Dreams: Striking Out for a New Feminine Future* (Boston: Little, Brown and Company, 1991), p. 95.
28. Traweek, *Beamtimes and Lifetimes*, pp. 90–91.
29. Interview on February 4, 1991, with Kristina Katsaros.
30. Interview on February 13, 1991, with Cynthia Haggerty.
31. I'd like thank construction engineer Viki Sonntag for her insights about cooperation and competition in trade associations.
32. Broad, *Betrayers of Truth*.
33. Riane Eisler, *The Chalice and the Blade: Our History, Our Future* (San Francisco: Harper & Row, 1987), p. xvii.
34. Matina S. Horner, "Fail: Bright Women," *Psychology Today* 3 (1969): pp. 36–38.
35. Phillip Shaver, "Questions Concerning Fear of Success and Its Conceptual Relatives," *Sex Roles* 2 (1979): pp. 205–220.

36. Rosser, *Female-Friendly Science*, p. 69.
37. Interview on February 1, 1991, with Paula Szkody.
38. Interview on January 14, 1990, with Davida Y. Teller.
39. Sigma Xi, *A New Agenda for Science*, p. 43.
40. Vera Kistiakowsky, "Women in Physics: Unnecessary, Injurious and Out of Place?" *Physics Today* 33 (February 1980): p. 32.
41. Interview on March 11, 1991, with her colleague, Becca Dickstein.
42. Interview on January 17, 1990, with Aimee Bakken.
43. From discussions at the weekly Women, Science, and Technology Reading and Discussion Group held at the University of Washington in Seattle during the autumn of 1990.
44. Interview on February 4, 1991, with Kristina Katsaros.
45. Interview on January 14, 1990, with Davida Y. Teller.
46. Rajiv Gandhi, April 4, 1985. The Pacific Science Center in Seattle used this quote to open their exhibit on science in India.
47. R. Buckminster Fuller and Anwar Dil, *Humans in Universe* (New York: Mouton, 1983), p. 112.
48. Ibid., p. 159.
49. Ibid., p. 112.
50. Ibid., p. 13.
51. R. Buckminister Fuller, *Intuition* (New York: Doubleday, 1970).

CHAPTER 9: INTUITION: ANOTHER WAY OF KNOWING

1. Mario Bunge, *Intuition and Science* (Westport, Conn.: Greenwood Press, 1962).
2. Frances E. Vaughan, *Awakening Intuition* (Garden City, N.Y.: Anchor Press, Doubleday, 1979).
3. Judith Hall, "Female Intuition Measured at Last?" *New Society* (London, 1977).
4. Interview on January 11, 1990, with Ingrith J. Deyrup-Olsen.
5. Mitroff, *The Subjective Side of Science*, p. 124.
6. Interview on March 27, 1991, with Sara Solla from the Research and Communication Division of AT&T Bell Laboratories in Holmdel, N.J.
7. Mitroff, *The Subjective Side of Science*, pp. 123–124.
8. C. G. Jung, *Psychological Types*, p. 453, ¶770.
9. Robert Teitelman, *Gene Dreams: Wall Street, Academia, and the Rise of Biotechnology* (New York: Basic Books, 1989).
10. Interview on February 13, 1991, with Cynthia Haggerty.
11. Joseph B. Wheelwright, *Saint George and the Dandelion: Forty Years of*

Practice as a Jungian Analyst (San Francisco: C. G. Jung Institute of San Francisco, 1982), pp. 67–68.

12. Interview on March 8, 1991, with Marsha Landolt.

13. Interview on January 22, 1991, with Eberhard K. Riedel.

14. Willis Harmon and Howard Rheingold, *Higher Creativity: Liberating the Unconscious for Breakthrough Insights* (Los Angeles: Jeremy P. Tarcher, 1984), pp. 24–28.

15. From discussions at the weekly Women, Science, and Technology Reading and Discussion Group held at the University of Washington in Seattle during the winter of 1991.

16. Shinichi Suzuki, *Nurtured by Love: A New Approach to Education* (Jericho, N.Y.: Exposition Press, 1969).

17. Quoted by Gerald Holton, *Thematic Origins of Scientific Thought: Kepler to Einstein*, revised ed. (Cambridge, Mass.: Harvard University Press, 1973, 1988), p. 305.

18. Quoted in Banesh Hoffmann and Helen Dukas, *Albert Einstein, Creator and Rebel* (New York: New American Library, 1973), p. 222.

19. Quoted by Holton, *Thematic Origins of Scientific Thought*, p. 368.

20. Freeman Dyson, *Disturbing the Universe* (New York: Basic Books, 1979), pp. 56–57.

21. Ibid., p. 62.

22. Ibid., pp. 54–55.

23. Letter on November 19, 1947, by Freeman Dyson in Richard P. Feynman (as told to Ralph Leighton), *"What Do You Care What Other People Think?": Further Adventures of a Curious Character* (New York: W. W. Norton & Co., 1988), p. 98.

24. Feynman, *"What Do You Care What Other People Think?,"* pp. 114–237.

25. Ibid., pp. 76–79.

26. Peter Medawar, *Pluto's Republic: Incorporating the Art of the Soluble and Induction and Intuition in Scientific Thought* (Oxford: Oxford University Press, 1982), p. 108.

27. A. Koestler, *The Act of Creation* (New York: Macmillan, 1964), p. 118.

28. Ibid.

29. L. Talamonti, *Forbidden Universe* (Briarcliff Manor, N.Y.: Stein and Day Publishers, 1975), p. 24.

30. James R. Newman, "Srinivasa Ramanujan," *Scientific American* 178 (June 1948): pp. 54–57.

31. B. M. Kedrov, "On the Question of Scientific Creativity," *Voprosy Psikologii* 3 (1957): pp. 91–113. Quoted by Willis Harmon and Howard Rheingold in *Higher Creativity*, pp. 30–31.

Notes

32. W. B. Kaempffert, *A Popular History of American Invention*, vol. 2 (New York: Scribner's, 1924).
33. Otto Loewi, "An Autobiographical Sketch," *Perspectives in Biology and Medicine* (Autumn 1960).
34. Ann Gibbons, "The Salk Institute at a Crossroads," *Science* 249 (27 July 1990): p. 360.
35. George Johnson, "Jonas Salk: May the 'Force' Be with Him," *Seattle Post-Intelligencer* (25 November 1990): pp. D1, D4.
36. C. G. Jung, *Psychology and Religion*, vol. 11 in *The Collected Works of C. G. Jung*, translated by R. F. C. Hull (Princeton: Princeton University Press, 1958), p. 12, ¶16.
37. Dennis Rawlins describes the debacle in "Starbaby," *Fate* 34 (October 1981): pp. 67–98.
38. Russell Targ and Keith Harary, *The Mind Race: Understanding and Using Psychic Abilities* (New York: Villard Books, 1984), pp. 14–17.
39. Ibid., p. 5.
40. Ibid., pp. 41–52.
41. Ibid., pp. 56–64.

CHAPTER 10. RELATEDNESS: A VISION OF WHOLENESS

1. Gabriele Uhlein, *Meditations with Hildegard of Bingen* (Santa Fe: Bear & Co., 1983), p. 41.
2. Edgar Mitchell, "Creating a New Reality," presented at the Thirteenth Annual Scientific Conference of the American Holistic Medical Association, March 31, 1990, in Seattle, Washington. Edgar Mitchell, Sc.D., who was one of the Apollo 14 astronauts and the sixth man to walk on the moon, is also a founder of both the Institute of Noetic Sciences and the Association of Space Explorers, and author of *Psychic Exploration: A Challenge for Science*.
3. Ibid.
4. Martin Lasden, "Closing in on Creation," *Stanford* (March 1990): p. 26.
5. Paul Davies, *The Cosmic Blueprint* (New York: Simon & Schuster, 1988), pp. 198–199. Davies is Professor of Mathematical Physics at the University of Adelaide in Australia.
6. Renée Weber, *Dialogues with Scientists and Sages: The Search for Unity* (London: Routledge & Kegan Paul, 1986), p. 29.
7. Interview on November 11, 1989, with Sylvia Pollack.
8. Ibid.
9. Interview on January 11, 1990, with Ingrith J. Deyrup-Olsen.
10. Barbara Sicherman and Carol Hurd Green, eds., *Notable American Women: The Modern Period* (Cambridge, Mass.: Belknap Press, 1980), p. 140.

11. Sicherman, *Notable American Women*, p. 140.
12. Uhlein, *Meditations with Hildegard of Bingen*, p. 10.
13. Constance Holden, "Multidisciplinary Look at a Finite World," *Science* 249 (July 6, 1990): pp. 18–19.
14. Ibid.
15. Kelley, *The Home Planet*, p. 71.
16. Interview on January 14, 1990, with Davida Y. Teller.
17. Interview on January 2, 1990, with Patricia Thomas.
18. Interview on January 17, 1990, Aimee Bakken.
19. Ibid.
20. Mitchell, "Creating a New Reality."
21. Gleick, *Chaos: Making a New Science*, pp. 174–175.
22. John Briggs, "Quantum Leap, an Interview with David Bohm," *New Age Journal* (September/October 1989): p. 49.
23. Ibid., p. 46.
24. John Briggs and F. David Peat, "Interview: David Bohm," *Omni* 9 (January 1987): p. 72.
25. Ibid.
26. Weber, *Dialogues with Scientists and Sages*, p. 51.
27. Albert Einstein, *Ideas and Opinions* (New York: Bonanza Books, 1954).
28. Briggs, "Quantum Leap," p. 49.
29. David Bohm, "Postmodern Science and a Postmodern World," in David Ray Griffin, ed., *The Reenchantment of Science* (New York: State University of New York Press, 1988), p. 67.

CHAPTER 11. THE SOCIAL RESPONSIBILITY OF SCIENCE

1. Ian Mitroff, *The Subjective Side of Science*, p. 114.
2. Ibid., p. 122.
3. Wolpert, *A Passion for Science*, p. 9.
4. Dyson, *Disturbing the Universe*, pp. 52–53.
5. Fuller, *Humans in Universe*, pp. 44, 48.
6. "A. Einstein: How I See the World."
7. Vare, *Mothers of Invention*, p. 147.
8. Interview on March 27, 1991, with Sara Solla.
9. Martha L. Crouch, "Confessions of a Botanist," *New Internationalist* (March 1991): p. 21.
10. Martha L. Crouch, "Debating the Responsibilities of Plant Scientists in the Decade of the Environment," *The Plant Cell* 2 (April 1990): pp. 275–277.

11. Interview with Marti Crouch on April 12, 1991.
12. Ibid.
13. Teri Klassen, "Scientist Gives Up Research She Says Hurts Environment," *Bloomington (Indiana) Herald-Times* (23 April 1990): pp. A1, A7.
14. Interview with Marti Crouch on April 12, 1991.
15. Ibid.
16. June I. Medford and Hector E. Flores, "Plant Scientists' Responsibilities: An Alternative," *The Plant Cell* 2 (June 1990): pp. 501–502.
17. Steven E. Smith, "Plant Biology and Social Responsibility," *The Plant Cell* 2 (May 1990): pp. 367–368.
18. Interview with Marti Crouch on April 12, 1991.
19. Gilligan, *In a Different Voice*, p. 21.
20. Ibid., p. 79.
21. National Academy of Sciences Committee on the Conduct of Science, *On Being a Scientist* (Washington, D.C.: National Academy Press, 1989), p. 20.
22. Interview on January 17, 1990, with Aimee Bakken.
23. Interview on March 8, 1991, with Marsha Landolt.
24. Ibid.
25. Interview with Sigrid Myrdal on November 11, 1989.
26. Ibid.
27. "A. Einstein: How I See the World."

CHAPTER 12. LIFTING THE VEIL: THE FEMININE IN EVERY SCIENTIST

1. Marie-Louise von Franz, *Shadow and Evil in Fairytales* (Dallas: Spring Publications, 1974), p. 69.
2. W. Ladson Hinton, "Some Darker Sides of the Human Soul," a paper given on February 3, 1992, at the C. G. Jung Society, Seattle.
3. Jung, *Psychological Types*, pp. 57–58.
4. Irene Claremont de Castillejo, *Knowing Woman: A Feminine Psychology* (New York: Harper & Row, 1973), p. 42.
5. These studies have been documented by Don Weitz in "A Psychiatric Holocaust: Canadian and CIA-sponsored Brainwashing Experiments," *Science for the People* (March/April 1987): pp. 13–19.
6. Abraham H. Maslow, *The Psychology of Science: A Reconnaissance* (New York: Harper & Row, 1966), pp. 30–31.
7. Merchant, *The Death of Nature*, p. 138.
8. From discussions at the weekly Women, Science, and Technology Read-

ing and Discussion Group held at the University of Washington in Seattle during the winter of 1991.

9. From Charles M. Johnston's lecture "Necessary Wisdom: The Challenge of a New Cultural Maturity" given to the C. G. Jung Society in Seattle on April 27, 1992.

10. Von Franz, *Shadow and Evil in Fairytales*, p. 116.

11. Interview on February 4, 1991, with Kristina Katsaros.

12. Interview on January 22, 1991, with Eberhard K. Riedel.

13. W. B. Yeats, "Under the Round Tower," in Richard J. Finneran, ed., *The Collected Poems of W. B. Yeats* (New York: Collier Books, Macmillan Publishing Co., 1989), p. 137.

BIBLIOGRAPHY

Achterberg, Jeanne. *Woman as Healer*. Boston: Shambhala Publications, 1990.

Appleman, Philip, ed. *Darwin: A Norton Critical Edition*. New York: Norton, 1970.

Aristotle. "On the Generation of Animals," in *The Works of Aristotle*, translated by Arthur Platt from vol. 2 of *The Great Books of the Western World*. Chicago: William Benton, publisher for Encyclopaedia Britannica, 1952.

Asch, Solomon E. "Opinions and Social Pressure." *Scientific American* 193, no. 5 (November 1955): 31–35.

Augros, Robert, and George Stanciu. *The New Biology: Discovering the Wisdom in Nature*. Boston: Shambhala Publications, New Science Library, 1988.

Axelrod, Robert, and William D. Hamilton. "The Evolution of Cooperation." *Science* 211 (27 March 1981): 1391.

Bacon, Francis. *Novum Organum* and *Of the Dignity and Advancement of Learning*, in J. Spedding, R. L. Ellis, and D. N. Heath, eds., *The Works of Francis Bacon*. London, 1858–61; reprinted Stuttgart: Friedrich Frommann Verlag, 1963.

Barker, E., trans. *The Politics of Aristotle*. Oxford: Oxford University Press, 1946.

Bass, N. H., A. Hess, A. Pope, and C. Thalheimer. "Quantitative Cytoarchitectonic Distribution of Neurons, Glia and DNA in Rat Cerebral Cortex." *Journal of Comparative Neurology* 143: 481–490.

Belenky, Mary Field, Blythe McVicker Clinchy, Nancy Rule Goldberger, and Jill Mattuck Tarule. *Women's Ways of Knowing: The Development of Self, Voice, and Mind*. New York: Basic Books, 1986.

Bellow, Saul, ed. *The Frontiers of Knowledge*. Garden City, N.Y.: Doubleday and Co., 1975.

Birke, Lynda, Wendy Faulkner, Sandy Best, Deirdre Janson-Smith, and Kathy Overfield, eds. *Alice Through the Microscope: The Power of Science over Women's Lives*. London: Virago, 1980.

Bibliography

Blakemore, Colin, and Grahame F. Cooper. "Development of the Brain Depends on the Visual Environment." *Nature* 228 (31 October 1970): 477–478.

Bleier, Ruth, ed. *Feminist Approaches to Science.* New York: Pergamon Press, 1988.

Bodger, Carole. "Salary Survey: Who Does What and for How Much?" *Working Woman* (January 1985): 72.

Bohm, David. *Wholeness and the Implicate Order.* London: Ark Paperbacks, 1980.

————, and F. David Peat. *Science, Order, and Creativity.* Toronto: Bantam Books, 1987.

Booth, William. "Oh, I Thought You Were a Man." *Science* 243 (27 January 1989): 475.

Born, Max. *Physics in My Generation.* London and New York: Pergamon Press, 1956.

Boslough, John. *Stephen Hawking's Universe: An Introduction to the Most Remarkable Scientist of Our Time.* New York: Avon Books, 1985.

Bowles, G., and R. Duelli-Klein, eds. *Theories of Women's Studies.* Boston: Routledge & Kegan Paul, 1983.

Briggs, John. "Quantum Leap, an Interview with David Bohm." *New Age Journal* (September/October 1989).

————, and F. David Peat. "Interview: David Bohm." *Omni* 9 (January 1987): 69–74.

————, and F. David Peat. *Looking Glass Universe: The Emerging Science of Wholeness.* New York: Simon and Schuster, 1984.

————, and F. David Peat. *Turbulent Mirror: An Illustrated Guide to Chaos Theory and the Science of Wholeness.* New York: Harper & Row, 1989.

Broad, William J. "Imbroglio at Yale (1): Emergence of a Fraud." *Science* 210 (1980): 38–41.

Broad, William, and Nicholas Wade. *Betrayers of the Truth.* New York: Simon & Schuster, 1982.

Bunge, Mario. *Intuition and Science.* Westport, Conn.: Greenwood Press, 1962.

Capra, Fritjof. *The Turning Point: Science, Society, and the Rising Culture.* Toronto: Bantam Books, 1982.

Chubin, Daryl E., and Edward J. Hackett. *Peerless Science: Peer Review and U.S. Science Policy.* Albany: State University of New York Press, 1990.

Cohn, Carol. "Sex and Death in the Rational World of Defense Intellectuals." *Signs: Journal of Women in Culture and Society* 12, no. 4 (1987): 687–718.

Colgrave, Sukie. *Uniting Heaven and Earth: A Jungian and Taoist Exploration of the Masculine and Feminine in Human Consciousness.* Los Angeles: Jeremy P. Tarcher, 1979.

Bibliography

Colinvaux, Paul. *Why Big Fierce Animals Are Rare: An Ecologist's Perspective.* Princeton: Princeton University Press, 1978.

Cooper, W. A., and E. N. Walker. *Getting the Measure of the Stars.* Bristol: Adam Hilger, 1989.

Cowen, R. "President's Budget: Rosy Outlook for R&D." *Science News* 139, no. 6 (9 February 1991): 87–94.

Crouch, Martha L. "Confessions of a Botanist." *New Internationalist* (March 1991): 21.

————. "Debating the Responsibilities of Plant Scientists in the Decade of the Environment." *The Plant Cell* 2 (April 1990): 275–277.

Darwin, Charles. *The Origin of Species,* 6th ed. London, 1872; reprinted New York: Mentor, 1958.

Davies, Paul. *The Cosmic Blueprint.* New York: Simon and Schuster, 1988.

Dawkins, R. *The Selfish Gene.* New York: Oxford University Press, 1976.

de Castillejo, Irene Claremont. *Knowing Woman: A Feminine Psychology.* New York: Harper & Row, 1973.

Diamond, Marian C., Arnold B. Scheibel, Greer M. Murphy, Jr., and Thomas Harvey. "On the Brain of a Scientist: Albert Einstein." *Experimental Neurology* 88 (1985): 198–204.

Dusheck, Jennie. "Female Primatologists Confer—Without Men." *Science* 249 (28 September 1990): 1494–1495.

Dyson, Freeman. *Disturbing the Universe.* New York: Basic Books, 1979.

Easlea, Brian. *Witch Craft, Magic and the New Philosophy.* Atlantic Highlands, N.J.: Humanities Press, 1980.

Easton, R., ed. "The World's Cats II." Seattle, Wash.: Feline Research Group, Woodland Park Zoo, 1976.

Ebel, Robert L. "What Is the Scientific Attitude?" *Science Education* 22, no. 2 (February 1938): 75–81.

Ehrenfeld, David. "The Next Environmental Crisis." *Conservation Biology* 3, no. 1 (March 1989): 1–3.

Einstein, Albert. *Ideas and Opinions.* New York: Bonanza Books, 1954.

Eisler, Riane. *The Chalice and the Blade: Our History, Our Future.* San Francisco: Harper & Row, 1987.

Eoyang, Eugene. "Chaos Misread: Or, There's a Wonton in My Soup." *Comparative Literature Studies* 26, no. 3 (1989): 271–284.

Ezzell, C. "Helping Cancers Mature So They Might Die." *Science News* 139 (1 June 1991): 341.

Farrington, B. "Thoughts and Conclusions." *The Philosophy of Francis Bacon.* Liverpool University Press, 1970.

Fausto-Sterling, Anne. *Myths of Gender.* New York: Basic Books, 1985.

Feynman, Richard P., as told to Ralph Leighton, *"What Do You Care What Other People Think?": Further Adventures of a Curious Character.* New York: W. W. Norton & Co., 1988.

Bibliography

Fowler, Charles, and Tim Smith, eds. *Dynamics of Large Mammal Populations.* New York: Wiley, 1981.
Fuller, Buckminster, and Anwar Dil. *Humans in Universe.* New York: Mouton, 1983.
Gallo, Robert. *Virus Hunting: AIDS, Cancer, and the Human Retrovirus: A Story of Scientific Discovery.* New York: Basic Books, 1991.
Galloway, Alison. "All Women Conference: Did It Discriminate?" *Science* 249 (7 December 1990): 1319.
Gibbons, Ann. "The Salk Institute at a Crossroads." *Science* 249 (27 July 1990): 360.
Gilligan, Carol. "The Conquistador and the Dark Continent: Reflections on the Psychology of Love." *Daedalus* 113 (1984): 75–95.
─────. *In a Different Voice: Psychological Theory and Women's Development.* Cambridge, Mass.: Harvard University Press, 1982.
Gillispie, Charles Coulston. *The Edge of Objectivity: An Essay in the History of Scientific Ideas.* Princeton: Princeton University Press, 1960.
Glanvill, Joseph. *The Vanity of Dogmatizing.* New York: Columbia University Press, 1931 (reproduced for the Facsimile Text Society), [1661].
Gleick, James. *Chaos: Making a New Science.* New York: Viking, 1987.
Goldman, Martin, and Marian Gordon Goldman. "Will She Make It?" *Working Woman* 9 (January 1984): 104.
Goodall, Jane van Lawick. *In the Shadow of Man.* Boston: Houghton Mifflin Company, 1971.
Gordon, Suzanne. *Prisoners of Men's Dreams: Striking Out for a New Feminine Future.* Boston: Little, Brown and Company, 1991.
Gribbin, John. *In Search of Schrödinger's Cat: Quantum Physics and Reality.* Toronto: Bantam Books, 1984.
Griffin, David Ray, ed. *The Reenchantment of Science.* New York: State University of New York Press, 1988.
Griffith-Marriott, Nancy. "Bodymind: An Interview with Candace Pert on Science, Feminism, Spirituality, and AIDS." *Woman of Power* 11 (Fall 1988): 22–25.
Hall, Judith. "Female Intuition Measured at Last?" *New Society* (London, 1977).
Hamilton, David P. "Publishing by—and for?—the Numbers." *Science* 250 (7 December 1990): 1331–1332.
Harding, M. Esther. *The Way of All Women.* New York: Harper & Row, 1970.
─────. *Woman's Mysteries: Ancient and Modern.* New York: Harper & Row, 1971.
Harding, Sandra, and Merrill B. Hintikka, eds. *Discovering Reality: Feminist Perspectives on Epistemology, Metaphysics, Methodology, and Philosophy of Science.* Boston: D. Reidel Publishing Company, 1983.
Harlow, Harry F. *Learning to Love.* New York: Jason Aronson, 1974.

Harmon, Willis, and Howard Rheingold. *Higher Creativity: Liberating the Unconscious for Breakthrough Insights.* Los Angeles: Jeremy P. Tarcher, 1984.

Hayles, N. Katherine. *Chaos and Disorder: Complex Dynamics in Literature and Science.* Chicago: University of Chicago Press, 1991.

Heisenberg, Werner. *Physics and Philosophy.* New York: Harper & Row, 1958.

Hirsch, Helmut V. B., and D. N. Spinelli. "Visual Experience Modifies Distribution of Horizontally and Vertically Oriented Receptive Fields in Cats." *Science* 168 (15 May 1970): 869–871.

Hoffmann, Banesh, and Helen Dukas. *Albert Einstein, Creator and Rebel.* New York: New American Library, 1973.

Holden, Constance. "Multidisciplinary Look at a Finite World." *Science* 249 (6 July 1990): 18–19.

Holton, Gerald. *Thematic Origins of Scientific Thought: Kepler to Einstein.* Cambridge, Mass.: Harvard University Press, 1973, 1988.

Horner, Matina S. "Fail: Bright Women." *Psychology Today* 3 (1969): 36–38.

Hubbard, Ruth. *The Politics of Women's Biology.* New Brunswick, N.J.: Rutgers University Press, 1990.

————, Mary Sue Henifin, and Barbara Fried, eds. *Biological Woman— The Convenient Myth.* Rochester, Vt.: Schenkman Books, 1982.

Jaynes, Julian. *The Origin of Consciousness and the Breakdown of the Bicameral Mind.* Boston: Houghton Mifflin, 1976.

Johnson, George. "Jonas Salk: May the 'Force' Be with Him." *Seattle Post-Intelligencer* (25 November 1990): D1, D4.

Johnson, J. T. "Fuzzy Logic." *Popular Science* (July 1990): 87–89.

Johnston, Charles. *Necessary Wisdom: Meeting the Challenge of a New Cultural Maturity.* Seattle: ICD Press, 1991.

Josephson, Brian. Letter to the editor. *Nature* 293 (15 October 1981): 594.

Jung, C. G. *The Archetypes and the Collective Unconscious,* vol. 9 in *The Collected Works of C. G. Jung.* Translated by R. F. C. Hull. Princeton: Princeton University Press, 1959.

————. *On the Nature of the Psyche,* from vol. 8 in *The Collected Works of C. G. Jung.* Translated by R. F. C. Hull. Princeton: Princeton University Press, 1971.

————. *Psychological Types,* vol. 6 in *The Collected Works of C. G. Jung.* A revision by R. F. C. Hull of the translation by H. G. Baynes. Princeton: Princeton University Press, 1971.

————. *Psychology and Alchemy,* vol. 12 in *The Collected Works of C. G. Jung.* Translated by R. F. C. Hull. Princeton: Princeton University Press, 1953.

————. *Psychology and Religion,* vol. 11 in *The Collected Works of C. G. Jung.* Translated by R. F. C. Hull. Princeton: Princeton University Press, 1958.

Bibliography

Jung, Emma. *Animus and Anima*. Dallas: Spring Publications, 1981.
Kaempffert, W. B. *A Popular History of American Invention*, vol. 2. New York: Scribner's, 1924.
Kass-Simon, G., and Patricia Farnes. *Women of Science: Righting the Record*. Bloomington, Ind.: Indiana University Press, 1990.
Kedrov, B. M. "On the Question of Scientific Creativity." *Voprosy Psikologii* 3 (1957): 91–113.
Keller, Evelyn Fox. *A Feeling for the Organism: The Life and Work of Barbara McClintock*. San Francisco: W. H. Freeman, 1983.
————. *Reflections on Gender and Science*. New Haven, Conn.: Yale University Press, 1985.
Kelley, Kevin W., ed. *The Home Planet*. Reading, Mass.: Addison-Wesley Publishing Company, 1988.
Kirk, David, ed. *Biology Today*. New York: Random House, 1975.
Kistiakowsky, Vera. "Women in Physics: Unnecessary, Injurious and Out of Place?" in *Physics Today* 33, no. 2 (February 1980): 32–40.
Klassen, Teri. "Scientist Gives Up Research She Says Hurts Environment." *The Bloomington (Indiana) Herald-Times* (23 April 1990): A1, A7.
Koestler, A. *The Act of Creation*. New York: Macmillan, 1964.
Koshland, Daniel E., Jr., "Waste Not, Want Some." *Science* 252 (26 April 1991): 485.
Kuhn, Thomas S. *The Structure of Scientific Revolutions*. Chicago: University of Chicago Press, 1970.
Lampkin, Richard H., Jr. "Scientific Attitudes." *Science Education* 22, no. 7 (December 1938): 353–357.
Lasden, Martin. "Closing in on Creation." *Stanford* (March 1990): 26.
Loewi, Otto. "An Autobiographical Sketch." *Perspectives in Biology and Medicine (Autumn 1960)*.
Lovelock, James E. "Small Science" in John Brockman, ed., *Doing Science: The Reality Club* (New York: Prentice Hall, 1988), p. 186.
Lowe, Marian, and Ruth Hubbard, eds. *Women's Nature: Rationalizations of Inequality*. Elmsford, N.Y.: Pergamon Press, 1983.
Luke, Helen. *Woman: Earth and Spirit, The Feminine in Symbol and Myth*. New York: Crossroad, 1981.
Maccoby, Eleanor, and Carol Nagy Jacklin. *The Psychology of Sex Differences*. Stanford, Calif.: Stanford University Press, 1974.
Macdonald, Sally. "Publish or Perish Throttles Teaching Role, Author Says." *The Seattle Times* (8 April 1990): B3.
Mandelbrot, Benoit B. *The Fractal Geometry of Nature*. New York: W. H. Freeman, 1977.
Mann, Charles. "Lynn Margulis: Science's Unruly Earth Mother." *Science* 252 (19 April 1991): 379.
Marco, Gino, Robert Hollingworth, and William Durham, eds. *Silent Spring Revisited*. Washington, D.C.: American Chemical Society, 1987.

Margulis, Lynn. *Symbiosis in Cell Evolution.* San Francisco: W. H. Freeman, 1981.

————. "Words as Battle Cries—Symbiogenesis and the New Field of Endocytobiology." *BioScience* 40 (October 1990): 675–676.

Martin, Brian. *The Bias of Science.* Canberra, Australia: Society for Social Responsibility in Science, 1979.

Maslow, Abraham H. *The Psychology of Science: A Reconnaissance.* New York: Harper & Row, 1966.

May, Robert M. "A Test of Ideas about Mutualism." *Nature* 307 (February 1984): 410.

McCormack, T. "Good Theory or Just Theory? Toward a Feminist Philosophy of Social Science." *Women's Studies International Quarterly* 4 (1981): 1–12.

Mead, Margaret. *Sex and Temperament in Three Primitive Societies.* New York: William Morrow, 1935.

Medawar, Peter. *Pluto's Republic: Incorporating the Art of the Soluble and Induction and Intuition in Scientific Thought.* Oxford: Oxford University Press, 1982.

Medford, June I., and Hector E. Flores. "Plant Scientists' Responsibilities: An Alternative." *The Plant Cell* 2 (June 1990): 501–502.

Merchant, Carolyn. *The Death of Nature: Women, Ecology, and the Scientific Revolution.* New York: Harper & Row, 1980.

Mitroff, Ian. *The Subjective Side of Science: A Philosophical Inquiry into the Psychology of the Apollo Moon Scientists.* Seaside, Calif.: Intersystems Publications, 1983.

Montgomery, Sy. *Walking with the Great Apes: Jane Goodall, Dian Fossey, Biruté Galdikas.* Boston: Houghton Mifflin Company, 1991.

Murdock, Maureen. *The Heroine's Journey.* Boston: Shambhala Publications, 1990.

Morison, Robert S. "Introduction." *Daedalus* 107, no. 2 (Spring 1978): vii–xvi.

National Academy of Sciences Committee on the Conduct of Science. "On Becoming a Scientist." Washington, D.C.: National Academy Press, 1989.

Neumann, Erich. *The Great Mother: An Analysis of the Archetype.* Translated by Ralph Manheim. Princeton: Princeton University Press, 1955.

————. *The Origins and History of Consciousness.* Translated by R. F. C. Hull. Princeton: Princeton University Press, 1954.

Newman, James R. "Srinivasa Ramanujan." *Scientific American* 178 (June 1948): 54–57.

Nicolis, Grégoire, and Ilya Prigogine. *Exploring Complexity: An Introduction.* New York: W. H. Freeman, 1989.

Noble, David F. "A World Without Women." *Technology Review* (May/June 1992): 53–60.

Overbye, Dennis. "Einstein in Love." *Time* (30 April 1990): 108.

Bibliography

Owen-Smith, Norman. "Territoriality in the White Rhinoceros (*Ceratotherium simium*) Burchell." *Nature* 231 (4 June 1971): 294.
Perry, William G. *Forms of Intellectual and Ethical Development in the College Years*. New York: Holt, Rinehart & Winston, 1970.
Polanyi, M. *The Logic of Liberty: Reflections and Rejoinders*. London: Routledge & Kegan Paul, 1951.
Prigogine, Ilya, and Isabelle Stengers. *Order Out of Chaos: Man's New Dialogue with Nature*. Toronto: Bantam Books, 1984.
Rawlins, Denis. "Starbaby." *Fate* 34 (October 1981): 67–98.
Roberts, Hilary. "A Qualified Failure." *New Scientist* 9 (June 1983): 722.
Roberts, Leslie. "The Rush to Publish." *Science* 251 (18 January 1991): 260–263.
Rose, H., and S. Rose, eds. *Ideology of/in the Natural Sciences*. Cambridge, Mass.: Schenkman, 1976.
Rosser, Sue V. *Female-Friendly Science: Applying Women's Studies Methods and Theories to Attract Students*. New York: Pergamon Press, 1990.
Rossiter, Margaret. *Women Scientists in America: Struggles and Strategies to 1940*. Baltimore: Johns Hopkins University Press, 1984.
Roth, Julius A. "Hired Hand Research." *The American Sociologist* (August, 1966): 190–196.
Schatten, G., and H. Schatten. "The Energetic Egg." *The Sciences* 23, no. 5 (1983): 28–34.
Scheffler, Israel. *Science and Subjectivity*. Indianapolis: Bobbs-Merrill Company, 1967.
Shaver, Phillip. "Questions Concerning Fear of Success and Its Conceptual Relatives." *Sex Roles* 2 (1979): 205–220.
Shiebinger, Londa. "The History and Philosophy of Women in Science: A Review Essay." *Signs: Journal of Women in Culture and Society* 12, no. 2 (1987): 305–332.
————. *The Mind Has No Sex*. Cambridge, Mass.: Harvard University Press, 1989.
Shrukin, Joel N. "Sexism and Hypocrisy." *Science* 249 (16 November 1990): 887.
Sicherman, Barbara, and Carol Hurd Green, eds. *Notable American Women: The Modern Period*. Cambridge, Mass.: Belknap Press, 1980.
Sigma Xi, the Scientific Research Society. *A New Agenda for Science*. New Haven, Conn.: Sigma Xi, 1986.
Singer, Charles, ed. *Studies in the History and Method of Science*. Oxford: Clarendon Press, 1921.
Sinsheimer, Robert L. "The Presumptions of Science." *Daedalus* 107, no. 2 (Spring 1978): 23–35.
Smith, Steven E. "Plant Biology and Social Responsibility." *The Plant Cell* 2 (May 1990): 367–368.
Stein, Murray. *In MidLife*. Dallas: Spring Publications, 1983.

Suzuki, Shinichi. *Nurtured by Love: A New Approach to Education.* Jericho, N.Y.: Exposition Press, 1969.

Talamonti, L. *Forbidden Universe.* Briarcliff Manor, N.Y.: Stein and Day Publishers, 1975.

Tannen, Deborah. *You Just Don't Understand: Women and Men in Conversation.* New York: William Morrow, 1990.

Targ, Russell, and Keith Harary. *The Mind Race: Understanding and Using Psychic Abilities.* New York: Villard Books, 1984.

Teitelman, Robert. *Gene Dreams: Wall Street, Academia, and the Rise of Biotechnology.* New York: Basic Books, 1989.

Traweek, Sharon. *Beamtimes and Lifetimes: The World of High Energy Physicists.* Cambridge, Mass.: Harvard University Press, 1988.

Tuana, Nancy, ed. *Feminism and Science.* Bloomington, Ind.: Indiana University Press, 1989.

Uhlein, Gabriele. *Meditations with Hildegard of Bingen.* Santa Fe: Bear & Co., 1983.

Ulanov, Ann Belford. *The Feminine in Jungian Psychology and in Christian Theology.* Evanston, Ill.: Northwestern University Press, 1971.

_____. *Receiving Woman: Studies in the Psychology and Theology of the Feminine.* Philadelphia: Westminster Press, 1981.

Vare, Ethlie Ann, and Gregg Ptacek. *Mothers of Invention: From the Bra to the Bomb, Forgotten Women and Their Unforgettable Ideas.* New York: William Morrow, 1988.

Vaughan, Frances E. *Awakening Intuition.* Garden City, N.Y.: Anchor Press, Doubleday, 1979.

Vetter, Betty M., and Eleanor L. Babco, eds. *Professional Women and Minorities: A Manpower Data Resource Service.* Commission of Professionals in Science and Technology, December 1987.

von Franz, Marie-Louise. *Shadow and Evil in Fairytales.* Dallas: Spring Publications, 1974.

von Senden, M. *Space and Sight: The Perception of Space and Shape in the Congenitally Blind Before and After Operation.* Translated by Peter Heath. Glencoe, Ill.: Free Press, 1960.

Watson, Lyall. *Beyond Supernature: A New Natural History of the Supernatural.* Toronto: Bantam Books, 1988.

Weber, Renée. *Dialogues with Scientists and Sages: The Search for Unity.* London: Routledge & Kegan Paul, 1986.

Weitz, Don. "A Psychiatric Holocaust: Canadian and CIA-sponsored Brainwashing Experiments." *Science for the People* (March/April 1987): 13–19.

Went, Frits W. "The Ecology of Desert Plants." *Scientific American* 192 (April 1955): 68–75.

_____. *The Plants.* New York: Time-Life Books, 1963.

Wheelwright, Joseph B. *Saint George and the Dandelion: Forty Years of*

Bibliography

Practice as a Jungian Analyst. San Francisco: C. G. Jung Institute of San Francisco, 1982.

Wilmore, Jack H. "Inferiority of Female Athletes, Myth or Reality?" *Journal of Sports Medicine* 3 (1975): 1–6.

Wilson, E. O. *On Human Nature.* Cambridge, Mass.: Harvard University Press, 1978.

Wolkomir, Richard. "Quark City." *Omni* 6, no. 5 (February 1984): 41.

Wolpert, Lewis, and Alison Richards. *A Passion for Science.* Oxford: Oxford University Press, 1988.

Young-Eisendrath, Polly, and Florence L. Wiedemann. *Female Authority: Empowering Women through Psychotherapy.* New York: Guilford Press, 1987.

Youth, Howard. "Iguana Farms, Antelope Ranches." *World Watch* 4, no. 1 (January/February 1991): 37–39.

Zweig, Connie, ed. *To Be a Woman: The Birth of the Conscious Feminine.* Los Angeles: Jeremy P. Tarcher, 1979.

INDEX

Traweek, Sharon, 100, 126–127,
137, 139–140, 156

Ulanov, Ann Belford, 33
Uncertainty
as freedom, 114–115
Heisenberg's uncertainty principle,
113, 114, 131

Vaughan, Frances, 203
Verner, Dagmar, 152
von Franz, Marie-Louise, 267–268,
281

Wade, Nicholas, 141
Walker, E. N., 149
Wallace, Alfred R., 107, 177
Wang, Taylor, 74–75
Watson, James, 60, 186
Wave/particle duality, 113, 129–
132
Weber, Renee, 247
Went, Frits, 178–179
Wheelwright, Joseph, 209–210

Wholeness
in chaos science, 95, 239–242
in integrated lives, 12, 246–248
in quantum theory, 242–246
vision of, 233, 235, 237
Wiedemann, Florence L., 15, 33
Wilson, E. O., 3, 48
Women
biology of, 41, 45–46
discrimination against, 5–6, 21, 23,
30, 41, 49, 147, 150–151, 195,
198–199
intellectual and moral development
of, 34–40, 191, 249
projection onto, 4, 278–279
See also Feminine
Women's Ways of Knowing, 34, 36–
40
World Wildlife Fund, 150
Worldview. *See* Reality

Yalow, Rosalyn, 6
Yin/Yang, 8–10, 13, 28, 226
Young-Eisendrath, Polly, 15

978-0-595-45771-7
0-595-45771-1

CPSIA information can be obtained at www.ICGtesting.com
Printed in the USA
LVOW13s2258050214

372571LV00001B/166/A